Lecture Notes in Civil Engineering

Volume 1

Lecture Notes in Civil Engineering (LNCE) publishes the latest developments in Civil Engineering - quickly, informally and in top quality. Though original research reported in proceedings and post-proceedings represents the core of LNCE, edited volumes of exceptionally high quality and interest may also be considered for publication. Volumes published in LNCE embrace all aspects and subfields of, as well as new challenges in, Civil Engineering. Topics in the series include:

- Construction and Structural Mechanics
- Building Materials
- Concrete, Steel and Timber Structures
- Geotechnical Engineering
- Earthquake Engineering
- Coastal Engineering
- Ocean and Offshore Engineering; Ships and Floating Structures
- Hydraulics, Hydrology and Water Resources Engineering
- Environmental Engineering and Sustainability
- Structural Health and Monitoring
- Surveying and Geographical Information Systems
- Indoor Environments
- Transportation and Traffic
- Risk Analysis
- Safety and Security

To submit a proposal or request further information, please contact the appropriate Springer Editor:

- Mr. Pierpaolo Riva at pierpaolo.riva@springer.com (Europe and Americas);
- Ms. Swati Meherishi at swati.meherishi@springer.com (Asia - except China - and Australia/NZ);
- Ms. Li Shen at li.shen@springer.com (China).

Indexed by Scopus

More information about this series at http://www.springer.com/series/15087

Paulo A. G. Piloto · João Paulo Rodrigues ·
Valdir Pignatta Silva
Editors

Advances in Fire Safety Engineering

Selected Papers from the 5th Iberian-Latin-American Congress on Fire Safety, CILASCI 5, July 15–17, 2019, Porto, Portugal

Springer

Editors
Paulo A. G. Piloto
Instituto Politécnico de Bragança
Bragança, Portugal

João Paulo Rodrigues
Universidade de Coimbra
Coimbra, Portugal

Valdir Pignatta Silva
Universidade de São Paulo
São Paulo, Brazil

ISSN 2366-2557 ISSN 2366-2565 (electronic)
Lecture Notes in Civil Engineering
ISBN 978-3-030-36239-3 ISBN 978-3-030-36240-9 (eBook)
https://doi.org/10.1007/978-3-030-36240-9

This Springer imprint is published by the registered company Springer Nature Switzerland AG
The registered company address is: Gewerbestrasse 11, 6330 Cham, Switzerland

Preface

This book gathers selected contributions presented during the 5th Iberian-Latin-American Congress on Fire Safety (CILASCI), held in Porto, Portugal, from 15–17 July 2019. The CILASCI is held once every two years, with the aim of disseminating scientific and technical knowledge in the field of fire safety, attracting different players involved in this area of knowledge.

The 5th Iberian-Latin-American Congress on Fire Safety reflected the new developments achieved in a wide range of application areas. The papers included in this book were selected out of 78 manuscripts (full papers), and five invited lectures, written and presented during six parallel sessions from researchers around the world (Algeria, Australia, Belgium, Brazil, China, Czech Republic, France, Hong Kong, Italy, Mozambique, Portugal, Spain, UK and USA).

The selected papers were peer-reviewed during and after the congress, and this selection process has resulted in new, extended, revised and full original versions covering the experimental analysis of materials and structures, the computational modelling of structures and materials, the fire events in special buildings and spaces, the architectural issues and evacuation topics for fire safety in buildings.

Fire safety has been advancing fast as a result of research, development and innovation worldwide; the new research programmes, the support of new skilled professionals and the existence of advanced training programmes in Fire Science Technology are not only expected to increase the safety level of people, buildings and products, but are also going to produce a positive impact in the economy of each country and society.

The editors gratefully acknowledge the members of the scientific committee and the experts who carried out the reviews of the manuscripts. They are also grateful to the organizing committee from the Polytechnic Institute of Bragança (IPB), the

Faculty of Engineering of the University of Porto (FEUP) and the School of Engineering (ISEP) from the Polytechnic Institute of Porto, and the support from the Luso-Brazilian Association for Fire Safety (ALBASCI) is also acknowledged.

Paulo A. G. Piloto
João Paulo Rodrigues
Valdir Pignatta Silva

Contents

Model Error for Calculating the Structural Reliability of Dowel Connections in Fire Situations

Auro Cândido Marcolan Júnior(✉) and Poliana Dias de Moraes

Universidade Federal de Santa Catarina, Florianópolis, Brazil
auro.marcolan@yahoo.com.br

Abstract. In fire situation, the thermomechanical evaluation of dowel connections can be performed by analytical models, which may present errors due to uncertainties in the representation of physical behavior through an idealized mathematical model. The objective of this paper is to evaluate and quantify the model error for the calculation of structural reliability in fire situation of wood-steel-wood aligned multiple dowels connections subjected to shear. The analytical model is based on the reduced cross section method. The physical properties and the geometrical configurations come from experimental research by other authors. The results indicate that the implemented model presented unfavorable safety results. The wood embedment failure mode, indicated by the model, is consistent with the failure mode presented in the experimental researches. The model for wood-steel-wood dowel connections, based on the reduced cross section method, can be used in a structural reliability analysis, considering that the model error is a random variable. The results indicate that the development and the evaluation of analytical models that better represent the failure time are necessary.

Keywords: Model error · Dowel connections · Timber structures · Fire

1 Introduction

Connections are generally the critical elements in the design of timber structures in normal and fire situation. The loads and the stresses are transferred between the structural elements and from the structure to the foundation through them. In the wood close to the dowels, stress concentrations arise and they can cause the connection failure. In a fire situation, the outer layers of timber elements are transformed into char and, due to the low thermal conductivity of the material, a thermal gradient arises inside the elements. The thermomechanical behavior of timber structures connection is complex due to the phenomenon involving wood combustion. During its combustion, heat and mass transfer occurs inside the connection [1–4]. The metal dowels have their mechanical properties altered and conduct heat inside the connection, causing softening and charring in the wood, reducing the load capacity of the connection [1]. Therefore, in fire situation, timber connections have an internal temperature gradient that does not have a simple analytical modeling [2, 3]. The effects of this phenomenon on the fire resistance time of timber connection are commonly evaluated by the reduced cross

P. A. G. Piloto et al. (Eds.): CILASCI 2019, LNCE 1, pp. 1–11, 2020.
https://doi.org/10.1007/978-3-030-36240-9_1

section method or by the reduced properties method [5], but can also be evaluated by numerical methods [1], as the heat-transfer finite element model combined with Johansen's theory [2] and the heat-transfer finite element model combined with a component model [3]. In Peng [6], for wood-steel-wood dowel connections, the most observed failure mode was wood embedment, with excessive elongation of the holes, caused by the reduction of embedment strength by temperature. Thus, the capacity load of this type of connection can be calculated by Johansen's theory in conjunction with the reduced cross section method.

Connections are subjected to uncertainties in: the material properties, the geometry and the loading. Despite the importance of the theme, there are few studies on probabilistic analysis of connections and timber elements in general, but can be mentioned the works of Köhler [7] in the reliability of timber structures and Jockwer *et al.* [8] in the reliability of timber connections with multiple dowel-type fastener. In fire situation, the thermomechanical evaluation of dowel connections can be performed by analytical models, which may present errors due to uncertainties in the representation of physical behavior through an idealized mathematical model [9]. Model error may represent a considerable portion of the uncertainties present in a timber connection in fire situation reliability analysis and should be investigated.

The objective of this paper is to evaluate and quantify the model error for the calculation of structural reliability, in fire situation, of wood-steel-wood multiple dowels shear connections, based on the reduced cross section method and the Johansen's theory. This paper originally was published at 5th Iberian-Latin American Congress on Fire Safety, CILASCI 5 [10]. In order to accomplish the objective, configurations of wood-steel-wood multiple dowels shear connections were analyzed using an analytical model based on the reduced cross section method [1]. The geometric configurations and the physical properties of the evaluated connections were obtained from the experimental research of Laplanche [11], Chuo [12], Peng [6] and Audebert *et al.* [4], which provided the failure time of the tested connections. Using this model, the evolution over time of the load capacity of the connection was determinate until the failure, which was considered when the applied load became greater than the load capacity of the connection.

2 Material and Study Methods

In this section, the load-bearing capacity of dowel timber connections, the dowel connections of timber structures in fire situation, the reduced cross section method, the model for the determination of the load capacity in fire situation and the model error for failure time prediction are presented.

2.1 Model for the Determination of the Load Capacity of the Connection in Fire Situation

In order to determine the load capacity of wood-steel-wood multiple dowels shear connections in fire situation an analytical model based on the Johansen's theory [13] in conjunction with the reduced cross section method is proposed. The basic premises of the model are: the failure modes considered are ductile; the outer layer of the timber elements is transformed into char and the interior keeps its mechanical resistance [1], the reduction of the cross section is given by the charring rate; the mechanical properties of the materials is considered rigid-plastic; the load capacity of the connections is a function of the wood embedment strength and the dowel yield moment strength; the mechanical properties of the dowel are not altered by the temperature; the load between the dowels is evenly distributed; the failure is considered when the applied load became greater than the load capacity of the connection. The load capacity of the connection is obtained by the Johansen's theory by the equations presented in Table 1.

In the analytical model, the wood embedment strength was obtained by Eq. 1 [14] and the dowel yield moment by Eq. 2 [15]:

$$f_h = 0.082 \cdot (1 - 0.01 \cdot d) \cdot \rho, \tag{1}$$

$$M_y = \frac{f_y \cdot d^3}{6}, \tag{2}$$

where f_h is the wood embedment resistance (N/mm^2), d is the dowel diameter (mm), ρ is the wood density (kg/m^3), M_y is the dowel yield moment (N.mm) e f_y is the yield strength of the steel dowel (N/mm^2).

2.2 Load-Bearing Capacity of Dowel Timber Connections

Johansen's theory [13] for determining the load-bearing capacity of dowel timber connection is based on the hypothesis that both wood and dowel behave as rigid-plastic materials. The dowel is considered as a beam supported by a wooden foundation. The load capacity of the connections is a function of the wood embedment strength and the dowel yield moment strength [15]. This theory does not take into account fragile failure modes [6].

The equations, based on Johansen's theory [13], for the determination of the load capacity per shear plane of wood-steel-wood dowel connections, are shown in the Table 1. In Mode I, embedment failure occurs; In Mode II, the failure is due to the embedment on the wood and the formation of a plastic hinge on the dowel. In Mode III, the failure occurs in the wood and plastic hinges form on the dowel; In Mode IV, the failure occurs by yielding of the steel plate. The load capacity per shear plane is given by the lowest value obtained among the 4 failure modes.

The load capacity of the connection (R_{ef}) is given by:

$$R_{ef} = n_{ef} \cdot R \cdot n_p, \tag{3}$$

where R is the load capacity of a dowel per shear plane, n_{ef} is the effective number of dowels and n_p is the number of shear planes.

The effective number of dowels (n_{ef}) is defined by Eq. 4, where n is the total number of dowels, a_1 is the distance between the aligned dowels and d is the dowel diameter [14].

$$n_{ef} = min\begin{cases} n \\ n^{0.9} \sqrt[4]{\frac{a_1}{13d}}, \end{cases} \quad (4)$$

Table 1. Shear plane capacity of wood-steel-wood connections [15].

Load capacity	Failure mode	
$R = f_h \cdot m \cdot d,$	(I)	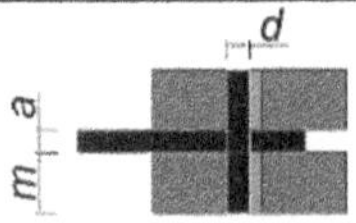
$R = f_h \cdot m \cdot d \left[\sqrt{2 + \frac{4M_y}{f_h \cdot d \cdot m^2}} \right],$	(II)	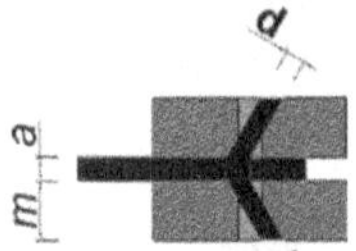
$R = \sqrt{2} \sqrt{2M_y \cdot f_h \cdot d},$	(III)	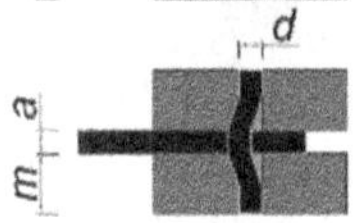
$R = [b - (d \cdot n_L)] \cdot a \cdot f_y,$	(IV)	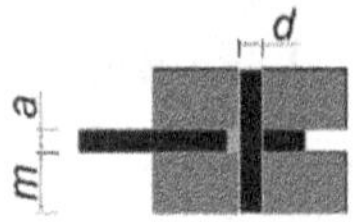

where R is the load capacity of a dowel per shear plane (N), f_h is the embedment strength (N/mm²), m is the thickness of the connection member (mm), d is the dowel diameter (mm), M_y is the dowel yield moment (N.mm), b is the width of the steel plate (mm), a is the thickness of the steel plate (mm) and n_L is the number of lines of dowels.

2.3 Reduced Cross Section Method

The reduced cross section method allows the calculation of the structural element resistance in a fire situation, taking into account the effective residual cross section of the element [1]. In this method, the mechanical properties of the residual cross section are assumed to be equal to that at room temperature. The reduction of cross section is considered by the charring rate [1]. The residual thickness, due to the charring of the wood-steel-wood connection section (Fig. 1), can be calculated by Eq. 5:

$$m_r = m - \beta \cdot t, \quad (5)$$

where m_r is the residual thickness of the wood elements in the connection at time t (mm), m is the original thickness of the wood elements in the connection (mm), β is the charring rate of the wood (mm/min), t is the fire exposure time (min) and m_{char} is the char thickness (mm). The charring rates used in the analytical model were those specified in Laplanche [11], Chuo [12], Peng [6] and Audebert *et al.* [4] for their tested connections.

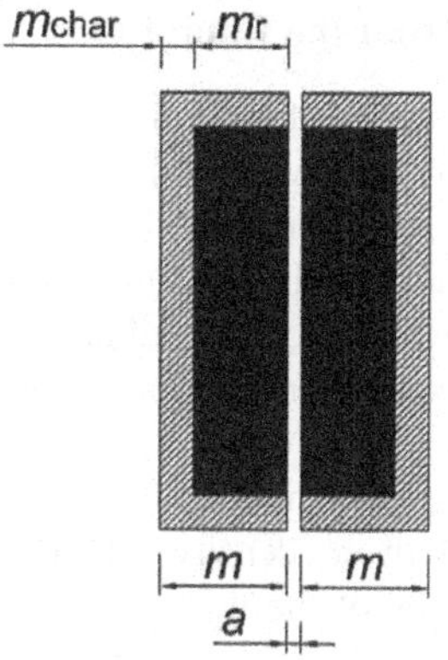

Fig. 1. Residual cross section and char layer of the wood-steel-wood connection.

2.4 Model Error for Failure Time Prediction

Model uncertainties are associated with the adopted simplifications in the model to represent a real phenomenon [9]. Simplifications of idealized models may not adequately represent the actual behavior of the phenomenon, since dowel connections of timber structures may exhibit stress concentrations and be subjected to unexpected behavior, especially in a fire situation. This generates uncertainties in the models associated with the adopted simplifications. For the evaluation of the model error, both the model and the phenomenon must provide as result the same variable. In this paper, the evaluation was performed by the time of failure in fire situation, that is the time when the maximum load capacity is reached.

Model error can be used to evaluate the quality of the model as a representation of the actual phenomenon and can be incorporated into the analysis as a random variable (X_m). This variable can be represented by a probability density function, or simply, by a mean and a standard deviation [16]. The model error can be calculated by Eq. 6:

$$\varepsilon_m = \frac{R_{exp}}{R_{teo}}, \tag{6}$$

where ε_m is the model error, R_{exp} is the experimentally obtained failure time and R_{teo} is the failure time obtained through the proposed model.

A model with a good prediction of the phenomenon has a mean proximally to a unit value and a coefficient of variation lesser than 10% [9]. The coefficient of variation is defined by the following equation:

$$C_V = \frac{\sigma_\varepsilon}{\mu_\varepsilon}, \tag{7}$$

where C_V is the coefficient of variation, σ_ε is the standard deviation of the model error, and μ_ε is the mean of the model error.

2.5 Experimental Results from the Literature

In this paper, 34 experimental results of multiple dowels double shear connections from the literature were used. The results were obtained from 4 distinct works: Laplanche [11], Chuo [12], Peng [6] and Audebert *et al.* [4]. In these researches, the connections were tested in a furnace, heated according to the ISO 834 standard fire curve [17]. The loads applied during the fire test were calculated taking into account the load at normal temperature test multiplied by a reduction factor (η) provided by the experimental studies [4, 6, 11, 12].

From Laplanche [11], 9 experimental results were obtained. The timber used was considered grade GL28h, with a density of 450 kg/m^3 and charring rate of 0.7 mm/min. The yield strengths of the dowel and steel plate are 300 and 355 MPa, respectively. The geometry data of the connections and the results of these tests are shown in Table 2.

From Chuo [12], 2 experimental results were obtained. The wood used was LVL, with a density of 650 kg/m^3 and charring rates of 0.8 and 0.9 mm/min. The yield strengths of the dowel and the steel plate are 320 and 275 MPa, respectively. The geometry data of the connections and the results of these tests are shown in Table 3.

From Peng [6], 13 experimental results were obtained. The wood used was considered grade 20f-EX, with a density of 455 kg/m^3 and charring rate of 0.8 mm/min. The yield strengths of the dowel and steel plate are 310 and 300 MPa, respectively. The geometry data of the connections and the results of these tests are shown in Table 4.

From Audebert *et al.* [4], 9 results were obtained. The wood used was grade GL28h with density of 450 kg/m^3 and charring rate of 0.7 mm/min. The yield strengths of the dowel and the steel plate are 336 and 360 MPa, respectively. The geometry data of the connections and the results of these tests are shown in Table 5.

Table 2. Results of tests performed by Laplanche [11].

Test	Normal temperature load (kN)	η	Fire test load (kN)	d (mm)	m (mm)	a (mm)	n_p	n_L	Failure time (min)
1	335	0.1	33	16	75	8	8	2	55.5
2	335	0.2	67	16	75	8	8	2	41
3	335	0.3	100	16	75	8	8	2	36
4	456	0.1	45	20	75	8	8	2	52
5	456	0.3	136	20	75	8	8	2	37
6	198	0.1	19	12	75	8	8	2	54
7	198	0.3	59	12	75	8	8	2	39
8	450	0.1	45	20	95	8	8	2	90
9	450	0.3	135	20	95	8	8	2	45

Table 3. Results of tests performed by Chuo [12].

Test	Normal temperature load (kN)	η	Fire test load (kN)	d (mm)	m (mm)	a (mm)	n_p	n_L	Failure time (min)
10	49	0.14	7	12	45	6	1	1	20
11	335	0.12	40	12	45	6	4	2	17.5

Table 4. Results of tests performed by Peng [6].

Test	Normal temperature load (kN)	η	Fire test load (kN)	d (mm)	m (mm)	a (mm)	n_p	n_L	Failure time (min)
12	58	0.1	6	13	38	9.5	2	1	14.5
13	58	0.1	6	13	38	9.5	2	1	15
14	58	0.29	16	13	38	9.5	2	1	8
15	58	0.3	17	13	38	9.5	2	1	8,5
16	115	0.11	12	13	60	9.5	4	2	28
17	115	0.18	20	13	60	9.5	4	2	22.5
18	115	0.29	33	13	60	9.5	4	2	17.5
19	65	0.11	7	19	60	9.5	1	1	27
20	65	0.32	20	19	60	9.5	1	1	15
21	227	0.1	22	19	60	9.5	4	2	26
22	227	0.3	68	19	60	9.5	4	2	14
23	243	0.1	24	19	80	9.5	4	2	36.5
24	243	0.29	70	19	80	9.5	4	2	19

Table 5. Results of tests performed by Audebert *et al.* [4].

Test	Normal temperature load (kN)	η	Fire test load (kN)	d (mm)	m (mm)	a (mm)	n_p	n_L	Failure time (min)
25	335	0.1	33	16	76	8	8	2	55.5
26	335	0.2	67	16	76	8	8	2	41
27	335	0.3	100	16	76	8	8	2	36
28	374	0.3	112	16	76	8	8	2	36
29	456	0.1	45	20	75	10	8	2	52
30	456	0.3	136	20	75	10	8	2	37
31	198	0.1	19	12	77	6	8	2	54
32	198	0.3	59	12	77	6	8	2	39
33	450	0.1	45	20	100	10	8	2	90
34	450	0.3	135	20	100	10	8	2	45

3 Results and Discussions

In this section the failure times of the model, the failure modes, the load capacity and the model error are shown and discussed.

3.1 Failure Times

The failure times obtained by the analytical model developed in this paper, for the different configurations of connections found in experimental tests reported in the literature was shown in Table 6. The failure times obtained by the analytical model were greater than the obtained in the experimental tests, which indicate that the model is not conservative and has an unfavorable behavior to safety.

3.2 Failure Mode

In this research the predominant failure mode was mode I, which is coherent with the experimental results found in the literature [4, 6, 11, 12], in which the failure occurs usually due to the wood embedment, with the elongation of the holes by the action of the dowels at high temperatures.

3.3 Load Capacity

The load capacity of the model and the experimental load capacity for the different configurations of connections found in experimental tests reported in the literature was shown in Table 6. The differences between the load capacity of the model and the experiments could be related to some limitations of the model. In which stands out that the model does not take into account the changes in the materials properties by the elevation of the temperature [2–4], the mechanical properties of the materials are considered linear, even that close to the failure stage the comportment could be non-linear and the heat transfer inside the member is not taken into account.

3.4 Model Error

In Fig. 2, the relations between experimental failure times and those provided by the developed model are shown. Points to the right of the ideal model curve are observed, showing an unfavorable behavior to safety. The obtained model error was 0.3936 and the standard deviation was 0.1238, which results a coefficient of variation of 0.3145.

Therefore, this model is not considered as a good approximation of the real phenomenon, since a model with a good prediction of the phenomenon would have a coefficient of variation of approximately 0.1 [16]. Nevertheless, the knowledge of the mean and the standard deviation of the model error allows the use of the evaluated model in a structural reliability analysis.

Table 6. Model results based on the reduced cross section method.

nº	$t_{f,teo}$ (min)	R_{exp} (kN)	R_{teo} (kN)	Failure mode	nº	$t_{f,teo}$ (min)	R_{exp} (kN)	R_{teo} (kN)	Failure mode
1	102	33	28	I	18	72	33	31	I
2	96	67	61	I	19	77	7	7	I
3	90	100	95	I	20	60	20	20	I
4	101	45	40	I	21	79	22	21	I
5	87	136	133	I	22	65	68	66	I
6	103	19	18	I	23	107	24	23	I
7	94	59	57	I	24	93	70	68	I
8	129	45	44	I	25	102	33	28	I
9	116	135	130	I	26	96	67	62	I
10	50	6	6	I	27	90	100	96	I
11	47	23	22	I	28	88	112	107	I
12	50	5	4	I	29	99	45	44	I
13	50	5	4	I	30	86	136	131	I
14	40	16	16	I	31	104	19	19	I
15	40	17	16	I	32	96	59	55	I
16	81	12	10	I	33	135	45	39	I
17	77	20	20	I	34	121	135	133	I

where $t_{f,teo}$ is the failure time of the model (min), R_{exp} experimental load capacity (kN) and R_{teo} load capacity of the model (kN).

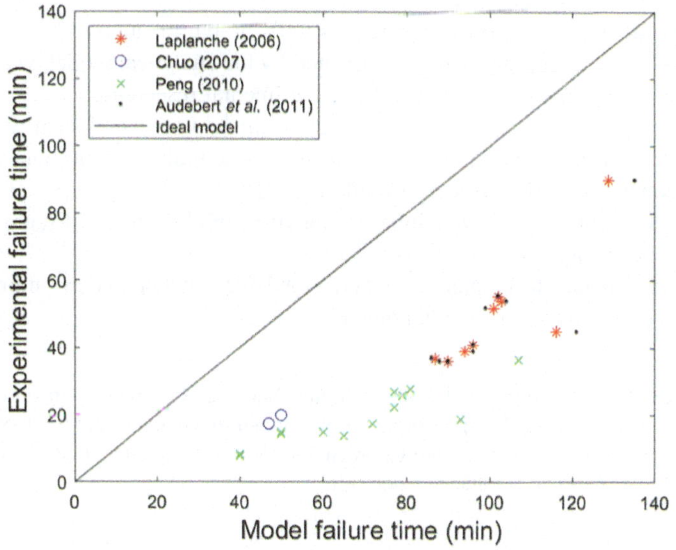

Fig. 2. Experimental versus model results by the reduced cross section method.

The model based on the reduced cross section method and the Johansen's equations [13] may be an alternative for calculating the structural reliability of wood-steel-wood dowels in fire situations, taking into account the model error calculated as random variable. As the model does not predict adequately the phenomenon, the development and the evaluation of analytical and numerical models that better represent the physical phenomenon of a wood connection in fire situation are recommended. In future works methods like the reduced properties method [1], heat-transfer finite element model combined with Johansen's theory [2] and the heat-transfer finite element model combined with a component model [3], could be evaluated by calculating their model error. By this a more faithful method to the phenomenon could be used to calculate the reliability of connections in fire situation. Model error can be a criterion in the selection of the model to be used in a reliability analysis. In addition, for a better understanding and statistical representation of the phenomenon, more experimental tests should be performed.

4 Conclusions

In this paper, an analytical model, based on the reduced cross section method and Johansen's theory, for the calculation of failure time of wood-steel-wood multiple dowels connections in fire situation was presented. From the comparison of the analytical results with the experimental ones [4, 6, 11, 12], calculation of the model error was possible. The results indicate that:

- the model error could be a quantitative criterion in the selection of the model to be used in a reliability analysis;
- the implemented model presented results unfavorable to safety;
- the wood embedment failure mode indicated by the analytical model, was consistent with the failure mode presented in the experimental researches;
- the model based on the reduced cross section method for wood-steel-wood dowels connections in fire situation can be used in a structural reliability analysis, taking into account the model error as a random variable;
- the development and the evaluation of analytical models that best represent failure time are necessary;
- more experimental tests should be performed for a better understanding and statistical representation of the phenomenon.

Acknowledgment. The authors thank the Conselho Nacional de Desenvolvimento Científico e Tecnológico – Brasil (CNPq) for sponsoring this research through grants number 148970/2016-8. This study was financed in part by the Coordenação de Aperfeiçoamento de Pessoal de Nível Superior – Brasil (CAPES) – Finance Code 001.

References

1. Buchanan, A.B., Abu, A.K.: Structural Design for Fire Safety, 2nd edn. Wiley, Chichester (2017)
2. Erchinger, C., Frangi, A., Fontana, M.: Fire design of steel-to-timber dowelled connections. Eng. Struct. **32**, 580–589 (2010)
3. Cachim, P., Franssen, J.: Numerical modelling of timber connections under fire loading using a component model. Fire Saf. J. **44**, 840–853 (2009)
4. Audebert, M., Dhima, D., Taazount, M., Bouchaïr, A.: Numerical investigations on the thermo-mechanical behavior of steel-to-timber joints exposed to fire. Eng. Struct. **33**, 3257–3268 (2011)
5. EN 1995-1-2: 2004: Design of timber structures – Part 1-2: General – Structural fire design. European Union (2004)
6. Peng, L.: Performance of heavy timber connections in fire. Doctoral Dissertation. Carleton University, Canada (2010)
7. Köhler, J.: Reliability of timber structures. Doctoral Dissertation. Swiss Federal Institute of Technology, Switzerland (2007)
8. Jockwer, R., Fink, G., Köhler, J.: Assessment of the failure behaviour and reliability of timber connections with multiple dowel-type fasteners. Eng. Struct. **172**, 76–84 (2018)
9. Melchers, R.E., Beck, A.T.: Structural Reliability Analysis and Prediction, 3rd edn. Wiley, Hoboken (2018)
10. Marcolan Jr., A.C., Moraes, P.D.: Erro de Modelo para o cálculo da Confiabilidade Estrutural de Ligações Parafusadas de Madeira em Situação de Incêndio. 5º Congresso Ibero-Latino-Americano em Segurança Contra Incêndio, Porto, pp. 553–562 (2019)
11. Laplanche, K.: Etude du comportement au feu des assemblages de structures bois: approche expérimentale et modélisation. Doctoral Dissertation. Université Blaise Pascal, France (2006)
12. Chuo, T.C.B.: Fire performance of connections in laminated veneer lumber. Doctoral Dissertation. University of Canterbury, New Zealand (2007)
13. Johansen, K.W.: Theory of Timber Connections. Int. Assoc. Bridg. Struct. Eng. **9**, 249–262 (1949)
14. EN 1995-1-1: 2004: Design of timber structures – Part 1-1: General – Common rules and rules for buildings. European Union (2004)
15. Thelandersson, S., Larsen, H.J.: Timber Engineering, 1st edn. Wiley, Chichester (2003)
16. Beck, A.T.: Confiabilidade e segurança das estruturas, 1st edn. Elsevier, Rio de Janeiro (2019)
17. ISO 834: Fire Resistance Testing. International Standards Organization (2000)

Three-Dimensional Numerical Analysis on the Fire Behaviour of Composite Slabs with Steel Deck

Paulo A. G. Piloto[1], Carlos Balsa[1(✉)], Fernando F. Ribeiro[2], and Ronaldo Rigobello[2]

[1] Instituto Politécnico de Bragança, Bragança, Portugal
balsa@ipb.pt
[2] Universidade Tecnológica Federal do Paraná, Campo Mourão, Brazil

Abstract. A composite steel-concrete slab consists of a concrete topping cast on the top of a profiled steel deck. Normally, the concrete is reinforced with an anti-crack mesh positioned on the upper part and individual reinforcing bars placed within the ribs. Composite slabs play an important role in the overall stability of buildings during a fire, and must be designed in accordance with regulations and standards. Usually, this structural element is rated on the basis of standard fire tests using the standard fire curve ISO 834. The fire resistance should be determined according to three different criteria, namely Load Bearing (R), Integrity (E) and Insulation (I). The Annex D of the EN 1994-1-2 presents a simplified calculation method for the determination of the fire resistance (I) of composite slabs subjected to standard fire exposure from below. During the last two decades, no revisions were made to this method, and there are no proposals for changes in the design formulae for the next version of the EN 1994-1-2. This investigation presents the development of numerical thermal models for three-dimensional analysis of composite slabs under fire conditions in ANSYS and MATLAB. During fire exposure, the steel deck heats up rapidly, expands and may separate from the concrete topping. In order to simulate debonding effects, an alternative thermal model is utilized, presenting an air gap with constant thickness between the steel deck and the concrete topping. The results of the numerical simulations are validated against the results of experimental fire tests, and compared with the simplified calculation method. A new equation is proposed for the calculation of the fire resistance (I) of composite slabs.

Keywords: Composite slabs · Fire resistance · Insulation criterion · Numerical simulation

1 Introduction

A composite steel-concrete slab consists of a concrete topping casted on the top of a profiled steel deck. Normally, the concrete is reinforced with an anti-crack mesh positioned on the upper part and individual reinforcing bars placed within the ribs, see Fig. 1. The steel deck acts as a permanent formwork and the composite action between the steel and concrete is generally achieved by indentations or embossments in the steel

P. A. G. Piloto et al. (Eds.): CILASCI 2019, LNCE 1, pp. 12–30, 2020.
https://doi.org/10.1007/978-3-030-36240-9_2

deck. Due to the reinforcement provided by the steel deck, composite slabs are generally slenderer and more efficient than flat full concrete slabs because they require less additional reinforcement and less concrete as well. The reduction of the construction time, simplicity of installation and reduction/elimination of shoring systems are other advantages of composite slabs that should be highlighted.

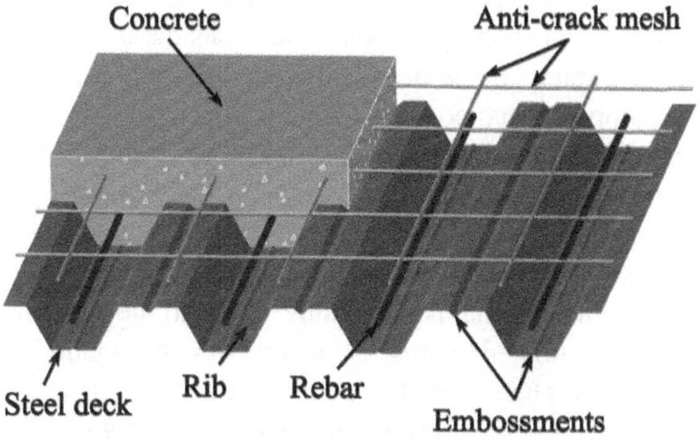

Fig. 1. Typical layout of a composite slab with trapezoidal steel deck.

Since the decade of 1980, a significant increase in the use of composite slabs with steel deck has taken place in Europe. The most popular types of shapes of the profiled steel deck are trapezoidal and re-entrant. Owing to the ease of casting concrete, slabs with trapezoidal steel deck are more popular than re-entrant ones. The overall depth of composite slabs usually varies between 100 and 170 mm, and the steel deck thickness between 0.7 and 1.2 mm. Generally, the steel deck is protected with a zinc layer on both faces in order to prevent corrosion and increase durability.

The steel deck may be directly exposed to accidental fire conditions. Composite slabs have to meet fire-safety requirements in accordance to standards and regulations. Normally, this structural element is rated on the basis of standard fire tests using the standard fire curve ISO 834 [1]. The fire resistance should be determined according to three different criteria, namely Load Bearing (R), Integrity (E) and Insulation (I).

The profiled geometry of the steel deck and the presence of the ribs in composite slabs create an orthotropic profile, resulting in complex thermal gradients hence presenting challenges in numerical modelling [2]. In recent years, several studies have been conducted in order to investigate the fire behaviour of this structural element. In 1983, recognizing the need for a calculation method, the European Convention for Constructional Steelwork (ECCS) [3] published the first instructions applied to the design of composite slabs with profiled steel deck under standard fire conditions. This document introduced simple calculation rules, which were based on the results of fire tests performed on different European laboratories, enabling the fire resistance of composite slabs to be quickly calculated. According to this technical note, for properly designed slabs at room temperature, the explicit fire analysis is not required to achieve a fire resistance of 30 min or less. In addition, it establishes that if the insulation criterion for fire resistance is fulfilled, then the integrity criterion is also fulfilled. At this time, the knowledge about the fire behaviour of composite slabs was incomplete and conservative assumptions were adopted, resulting in uneconomical solutions.

In 1991, Hamerlinck [4] conducted a numerical and experimental study regarding the thermal and mechanical behaviour of reinforced composite slabs under fire conditions. Both numerical models were experimentally validated with loaded and unloaded tests. Due to the melting of the zinc layer and surface blackening, the resulting emissivity of the galvanized steel deck was calculated as temperature dependent. The testing programme took into consideration the most important parameters for fire resistance and a new computer program was developed, enabling simulations at low computational cost (low time processing). It was concluded that the developed two-dimensional model provided satisfactory results, although not including three-dimensional thermal effects.

In 1998, an investigation was carried out by Both [5] with the main objective of introducing easy to handle calculation rules as well as providing more insight on the fire behaviour and failure mechanisms mainly of continuous composite slabs. The numerical models were validated against the results of experimental tests performed by the author and other researchers. The three-dimensional effects near internal supports, concrete cracking and the melting of the zinc layer of the steel deck were considered. Finally, a parametric study was performed and simple calculation rules were derived from the results using standard regression techniques. It was concluded that the thermal model was able to describe the two and three-dimensional heat flow in composite slabs during fire exposure and the assessment rules for the fire resistance given in Eurocode 4 at that time could be considerably improved, among other conclusions.

In 1999, Abdel-Halim, Hakmi and O'Leary [6] conducted a study with the main goal of providing data about the performance of fire exposed composite slabs adopting a model fire test facility. Two different specimens, one with and another without additional longitudinal reinforcement bars were tested using the standard fire ISO 834 in the University of Salford, UK. Thereupon, the investigation focused on the analysis of the effect of additional bars on the fire resistance as well as on the comparison between the fire resistance of the samples with respect to integrity and insulation criteria. It was concluded that the specimen without reinforcement bars presented a lower rate of temperature rise on the unexposed surface in comparison to the reinforced specimen and consequently, a higher fire resistance in both insulation and integrity criteria. This is explained due to the existence of high conductive material (steel rebar) within the concrete layer of the reinforced specimen.

In 2002, Lim and Wade [7] performed fire tests on six large-scale concrete slabs, comprising three reinforced concrete flat slabs and three composite steel-concrete slabs. The main objective of the tests was to analyse the fire behaviour of unrestrained simply supported slabs in a controlled furnace. The slabs were subjected to a live load and standard fire conditions during three hours. All the slabs resisted the full duration of the tests without collapsing, despite presenting extensive surface cracking on the unexposed surface and large deflections (up to 270 mm). In general, the measured fire resistance was higher than the predictions from normative recommendations. The results evidenced the important effect of membrane action on preserving the structural stability of the slabs under fire conditions.

In 2005, the European Commission for Standardization (CEN), published design rules for composite steel-concrete structures under fire conditions, EN 1994-1-2 [8]. This standard determines that if a composite slab with profiled steel deck, with or without additional reinforcement, is properly designed according to EN 1994-1-1 [9], the fire resistance (R) is at least 30 min. In addition, it also states that the integrity criterion (E) is always fulfilled for composite slabs with steel deck. Regarding the insulation criterion (I), the Annex D presents a simplified calculation method, which depends on the geometry of the steel deck, the thickness of concrete above the steel deck and the view factor of the upper flange.

In 2011, Guo and Bailey [10] executed an experimental investigation with the aim of providing more insight on the behaviour of composite slabs during heating and cooling phases of fire. Nine equal specimens were tested: two of them at room temperature and the others at three different fire scenarios, which were controlled by burners and fans within the furnace. The specimens were loaded with representative values found in practice in order to investigate the structural behaviour. The results showed that the maximum temperature and both heating and cooling rates strongly affected the behaviour of the slabs. For all the tests, the maximum temperatures on the unexposed surface and on the mesh were both higher during the cooling stages, due to the thermal inertial effect, which highlighted that insulation failure is likely to occur not only during heating but also during cooling phase.

In 2018, an investigation about the fire behaviour of composite slabs under standard fire conditions was carried out by Piloto et al. [11]. The key objective of this study was to develop two-dimensional numerical models using the software MATLAB and ANSYS in order to evaluate the fire resistance of different slab configurations according to the insulation criterion (I). Several numerical simulations were performed with the aim of analysing the effect of both concrete and steel deck thicknesses on the temperature development on the unexposed side. Two experimental fire tests were conducted on unloaded specimens. On the whole, the fire resistance (I) obtained from the numerical models was considerably smaller than those measured in the experimental tests. A comparison between the numerical results and the simplified calculation method of Eurocode 4 evidenced that the current design rules are unsafe, that is, the European standard overestimated the fire resistance in comparison to the numerical simulations made with perfect contact and underestimates the fire resistance with respect to the experimental test. According to the numerical results, a new and better approach considering a quadratic dependence between the fire resistance and the effective thickness was proposed.

In 2019, Jiang et al. [2], from the National Institute of Standards and Technology (NIST), conducted a numerical investigation around different parameters that may influence the fire resistance of composite slabs with respect to the thermal insulation criterion (I). An improved algebraic expression for the calculation of the fire resistance that explicitly accounts for moisture content of concrete was proposed. The formulation is applicable to an extended range of geometries in comparison to the limitations of the calculation method presented in the current version of Eurocode 4. A set of 54 composite slabs was selected for the numerical analysis, using an accurate finite element approach. It was concluded that the concrete thickness and the moisture content were the parameters that most influenced the fire resistance. The proposed expression for fire resistance was validated against additional analyses and experimental data, resulting in maximum deviations of 15 and 18 min, respectively.

The scope of this investigation concerns numerical simulations using the standard fire curve ISO 834 [1] in order to evaluate the fire resistance from the thermal insulation standpoint. Three-dimensional thermal models were implemented using ANSYS and MATLAB PDE toolbox.

During fire exposure, the steel deck heats up rapidly, expands and may separate from the concrete topping. Previous studies mention the separation between the steel deck and concrete during fire exposure, which increases the thermal resistance in this interface. In order to simulate the debonding effects, an alternative thermal model is presented, including an air gap with constant thickness between the steel deck and concrete topping.

With the aim of validating the approach, the results of the numerical simulations are compared with experimental results published by Lim and Wade [7], Piloto et al. [11], Abdel-Halim, Hakmi, O'Leary [6], and Hamerlinck [4]. In addition, a comparison between the fire resistance obtained numerically, experimentally and analytically (using the Eurocode 4 calculation method [8]) is presented.

This paper is divided into 7 sections. In Sect. 1, an introduction to the research theme and a summary of relevant investigations in the field of study are given. Section 2 presents the definition of the fire resistance criteria. Section 3 deals with the simplified calculation method of Eurocode 4. In Sect. 4, a succinct description of the experimental setup of the four different fire tests is given. Section 5 describes the numerical thermal model and presents the comparison of the results. Section 6 presents a parametric study and the proposed new equation for fire resistance (I). Finally, the conclusions and general observations about the results are given in Sect. 7.

2 Fire Resistance Criteria

Structural elements need to meet fire-safety requirements according to building codes and standards. For composite slabs, the requirements are normally specified by fire ratings of 30, 60, 90 min or more. The fire rating of this type of building elements is usually made using standard fire tests [12, 13] and should consider the criteria of Insulation (I), Integrity (E) and Load Bearing (R). Usually, experimental tests are expensive and time-consuming. As an alternative solution, the fire resistance can be determined by means of numerical simulations and simple calculation methods. The fire resistance of composite slabs is defined with respect to a standard fire exposure from below. In this study, the fire resistance is exclusively investigated with respect to the thermal insulation criterion (I).

The thermal insulation criterion (I) is the ability to withstand fire in one side and prevent excessive transmission of heat. The assessment shall be made on the basis of the average temperature rise on the unexposed surface limited to 140 °C above the initial average temperature, or; on the basis of the maximum temperature rise at any point on the unexposed surface limited to 180 °C above the initial average temperature.

The integrity criterion (E) is the capacity to withstand fire in one side and resist penetration of hot gases and flames. The assessment should be made on the basis of measuring cracks or openings in excess of given dimensions, or the ignition of a cotton pad, or sustained flaming on the unexposed side. For composite slabs cast in situ, the integrity criterion is normally fulfilled provided that the joints are adequately sealed.

The load bearing resistance for flexural loaded elements (R) is the ability to support the loading during the test without collapsing. The assessment shall be made on the basis of limiting vertical displacement D ($D = L^2/400d$ [mm]), or limiting the rate of vertical displacement ($dD/dt = L^2/9000d$ [mm/min]), being L the clear span of the testing specimen in millimetres and d the distance from the extreme fibre of the cold design compression zone to the extreme fibre of the cold design tensile zone of the structural section, in millimetres.

3 Simplified Calculation Method of Eurocode 4

The Annex D of EN 1994-1-2 [8] presents a simplified calculation method for the prediction of the fire resistance of unprotected composite slabs subjected to the standard fire curve ISO 834 from below. The analytical expressions given in the current version of this standard are based on the study conducted by Both [5] in 1998. During the last years, no revisions were made to this method [2], and there are no proposals for changes in the design formulae for the next version of the EN 1994-1-2. The fire resistance (t_i) with respect to thermal insulation criterion should be determined according to Eq. 1.

$$t_i = a_0 + a_1 \cdot h_1 + a_2 \cdot \phi_{upper} + a_3 \cdot \frac{A}{L_r} + a_4 \cdot \frac{1}{l_3} + a_5 \cdot \frac{A}{L_r} \cdot \frac{1}{l_3} \tag{1}$$

The rib geometry factor of the slab (A/L_r) shall be calculated according to Eq. 2.

$$A/L_r = h_2 \cdot ((l_1 + l_2)/2) \Big/ \left(l_2 + 2\sqrt{h_2^2 + ((l_1 - l_2)/2)^2} \right) \tag{2}$$

The fire resistance depends on the geometric parameters of the slab (l_1, l_2, l_3, h_1 and h_2), and also on partial factors (a_i). Table 1 presents the factors for composite slabs with normal weight concrete (NWC).

Table 1. Coefficients for determination of the fire resistance for composite slabs with NWC (adapted from EN 1994-1-2 [8]).

a_0 (min)	a_1 (min/mm)	a_2 (min)	a_3 (min/mm)	a_4 (mm min)	a_5 (min)
−28.8	1.55	−12.6	0.33	−735.0	48.0

The EN 1994-1-2 states that the effective thickness of a composite slab h_{eff} (mm) should be calculated according to Eq. 3a, 3b.

$$h_{eff} = h_1 + 0.5 \cdot h_2 \cdot \left(\frac{l_1 + l_2}{l_1 + l_3}\right) \quad \text{for } h_2/h_1 \leq 1.5, \text{and } h_1 > 40\,\text{mm} \tag{3a}$$

$$h_{eff} = h_1 \cdot \left[1 + 0.75 \cdot \left(\frac{l_1 + l_2}{l_1 + l_3}\right)\right] \quad \text{for } h_2/h_1 > 1.5, \text{and } h_1 > 40\,\text{mm} \tag{3b}$$

The geometric parameters of the slab h_1, h_2, l_1, l_2 and l_3 are illustrated in Figs. 2, 3, 4 and 5. The effective thickness may be adopted as h_1 if $l_3 > 2\ l_1$.

4 Experimental Fire Tests

Four different composite slabs with trapezoidal profile have been selected to perform the numerical validation. These slabs correspond to experimentally tested slabs: slab 1 was tested by Lim and Wade [7] (test number 4), slab 2 was tested by Piloto et al. [11] (test number 1), slab 3 was tested by Abdel-Halim, Hakmi and O'Leary [6] (test number 2), and slab 4 was tested by Hamerlinck [4] (test number 2). The slabs were exposed to the ISO 834 standard fire from below in controlled furnaces. The profiles of slabs 1, 2, 3 and 4 are shown, respectively, in Figs. 2, 3, 4 and 5.

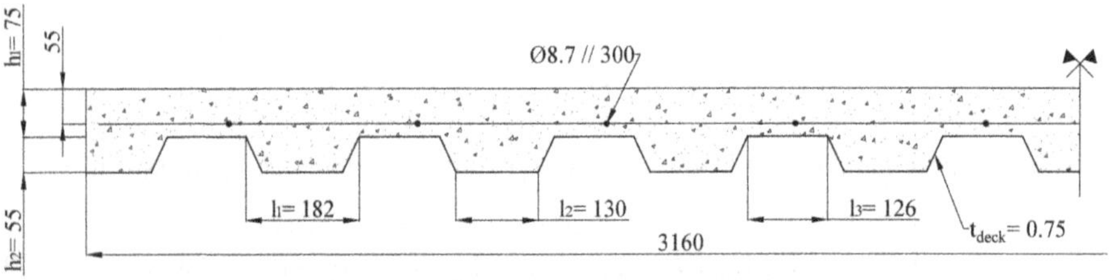

Fig. 2. Profile of slab 1: dimensions in millimetres (Lim and Wade [7]).

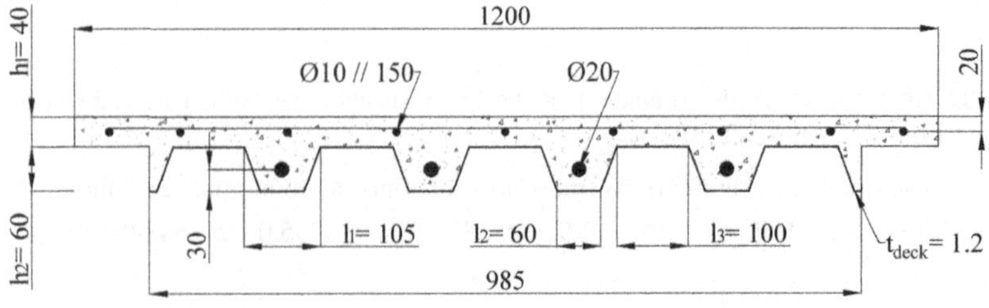

Fig. 3. Profile of slab 2: dimensions in millimetres (Piloto et al. [11]).

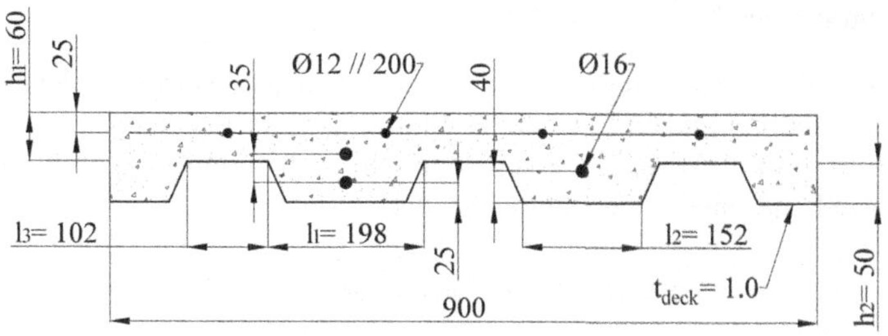

Fig. 4. Profile of slab 3: dimensions in millimetres (Abdel-Halim, Hakmi and O'Leary [6]).

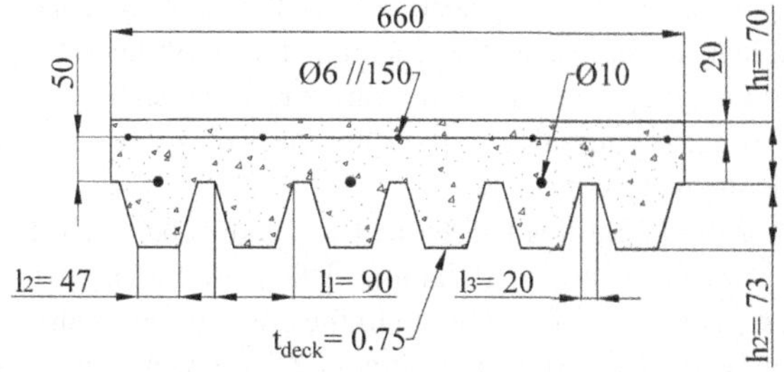

Fig. 5. Profile of slab 4: dimensions in millimetres (Hamerlinck [4]).

For slab 1 (clear span of 3160 mm wide by 4160 mm long), normal weight con crete with siliceous aggregates was used and the moisture content amounted to 5.6% by weight. The initial bulk temperature was of 13 °C. For slab 2 (clear span of 985 mm wide by 916.8 mm long), normal weight concrete was used and the moisture content was approximately 3.0% by weight. The initial bulk temperature amounted to 20 °C. For slab 3 (clear span of 900 mm wide by 1200 mm long), normal weight concrete was used and prior the test, the specimen was conditioned with the aim of reducing the moisture to air dry condition. The initial bulk temperature was of 20 °C. For slab 4 (clear span of 660 mm wide by 3200 mm long), normal weight concrete was used and the moisture content amounted to 3.5% by weight. The initial bulk temperature amounted to 20 °C. For all the slabs, debonding of the steel deck from the concrete topping was observed during the tests.

5 Numerical Modelling

In the following section, the methodology used to model and numerically determine the thermal effects of standard fire exposure on composite slabs is outlined. Thereby, a brief description of the finite elements, thermal properties of materials, boundary conditions and convergence criterion are presented. Finally, a comparison between the numerical, experimental and analytical results is presented.

5.1 Thermal Model

The composite slab is meshed to solve a nonlinear transient thermal analysis, using 3-D models from ANSYS and MATLAB. The finite element method (FEM) requires the solution of Eq. 4 in the domain and the definition of the boundary conditions in Eq. 5 on the exposed and unexposed side of the slab.

$$\nabla\left(\lambda_{(T)} \cdot \nabla T\right) = \rho_{(T)} \cdot Cp_{(T)} \cdot \partial T/\partial t \tag{4}$$

$$\lambda_{(T)} \cdot \nabla T \cdot \vec{n} = \alpha_c \cdot \left(T_g - T\right) + \phi \cdot \varepsilon_m \cdot \varepsilon_f \cdot \sigma \cdot \left(T_g^4 - T^4\right) \tag{5}$$

In the equations above: T represents the temperature of each material; $\rho_{(T)}$ is the specific mass; $Cp_{(T)}$ is the specific heat; $\lambda_{(T)}$ is the thermal conductivity; α_c is the convection coefficient. T_g represents the gas temperature of the fire compartment, using the standard fire ISO 834 applied on the bottom part of the slab; ϕ is the view factor; ε_m is the emissivity of each material; ε_f represents the emissivity of the fire and σ represents the Stefan-Boltzmann constant.

The view factor (ϕ) quantifies the geometric relation between the surface emitting radiation and the surface receiving radiation. The view factor of the lower flange of composite slabs (ϕ_{low}) is given as 1. Owing to the obstruction to direct exposure caused by the ribs of the steel deck, the view factors of the web (ϕ_{web}) and upper flange (ϕ_{upper}) are smaller than one. These view factors can be calculated as function of the geometric parameters of the slab, as follows.

$$\phi_{upper} = \frac{\sqrt{h_2^2 + \left(l_3 + \frac{l_1 - l_2}{2}\right)^2} - \sqrt{h_2^2 + \left(\frac{l_1 - l_2}{2}\right)^2}}{l_3} \tag{6}$$

$$\phi_{web} = \frac{\sqrt{h_2^2 + \left(\frac{l_1 - l_2}{2}\right)^2} + (l_3 + l_1 - l_2) - \sqrt{h_2^2 + \left(l_3 + \frac{l_1 - l_2}{2}\right)^2}}{2\sqrt{h_2^2 + \left(\frac{l_1 - l_2}{2}\right)^2}} \tag{7}$$

The finite element method is applied to solve numerically the heat transfer equation using the software ANSYS and MATLAB. For slab 1, the respective 3-D meshes developed in ANSYS and MATLAB are presented in Fig. 6.

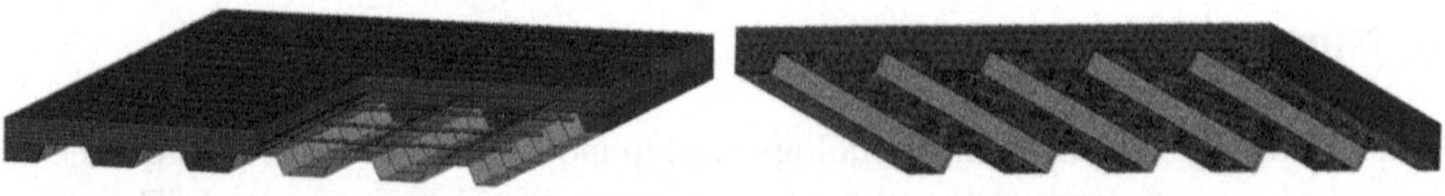

(a) ANSYS. (b) MATLAB.

Fig. 6. Finite element mesh (slab 1).

A 3-D model of the slab is generated, which is composed by subdomains that correspond to the different materials: the concrete topping, steel deck, rebars and steel mesh in ANSYS; and concrete topping and steel deck in MATLAB. A parametric study performed by the authors evidenced that the steel components within the concrete topping do not affect the fire resistance with respect to the thermal insulation criterion. Therefore, by means of simplification, these components are not included in the MATLAB numerical model. An alternative thermal model is created, using an air gap with constant thickness (1 mm for slab 1, 3 mm for slab 2, 2 mm for slab 3, and 1 mm for slab 4), included between the steel deck and the concrete topping in order to simulate the debonding effects. The thicknesses of the air layer used in the air gap model for the four slabs were determined from a parametric analysis, selecting the values that best fit with the experimental data for each slab. Through the air gap, only heat flow by conduction is considered, due to the very small thickness of this layer.

The thermal properties of the materials are temperature dependent and vary according to the standards used for composite slabs [8], steel structures [14] and concrete structures [15]. The thermal properties of steel, concrete and air are presented in Fig. 7. Presently, there is no standard which specifies the thermal properties of air. However, computer programs and experimental tests provide reliable data for numerical analyses. This work considers the thermal properties of air at 1 atm pressure [16]. Regarding the conductivity of concrete, the upper limit was selected for the numerical simulations. The specific heat of concrete presents a peak value related to 3% of moisture content of concrete weight. The extrapolation method was used to update higher moisture contents.

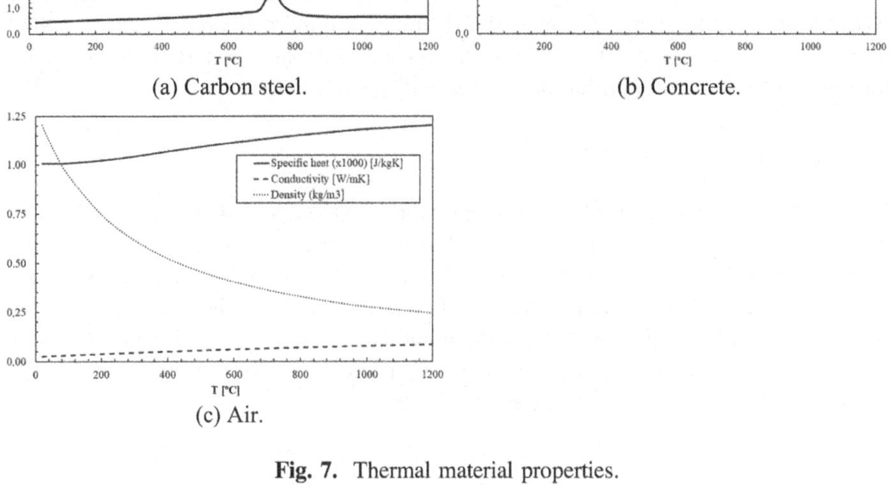

(a) Carbon steel.

(b) Concrete.

(c) Air.

Fig. 7. Thermal material properties.

In ANSYS, three different finite elements are used: SHELL131, SOLID70 and LINK33. The SHELL131 element has four nodes with up to 32 degrees of freedom (temperature) per node, depending on the number of layers (one layer). This element presents linear interpolation functions in the plane of the element, using full Gauss integration method (2 × 2) and linear interpolation functions through the layer thickness (three Gauss points). The shell element is used to model the steel deck of the composite slab. The bottom temperature of shell element nodes is assumed to be equal to the temperature of solid element nodes, when both nodes are coincident. The SOLID70 element presents eight nodes with a single degree of freedom (temperature) at each node. Linear interpolation functions are used for this element and the full Gauss integration method is also applied (2 × 2 × 2). This finite element is used to model the concrete topping and, in the alternative model, the air gap volume. The LINK33 element has two nodes with a single degree of freedom (temperature) per node. This uniaxial element presents linear interpolation functions and exact integration. The LINK33 element is used to model the anti-crack mesh and the rebars.

In MATLAB, the geometry is meshed in the Partial Differential Equation Toolbox (PDE) with only linear tetrahedral (solid) elements. The toolbox does not support meshes with elements of different types. This element has four nodes (one node at each corner) and a single degree of freedom (temperature) at each node. Linear interpolation functions are used for this element.

According to the measured initial average temperatures, see Sect. 4, all the nodes are set with an initial condition for temperature of 13 °C on slab 1, and with 20 °C on slabs 2, 3 and 4. The exposed side is submitted to a heat flux by convection and radiation, see Eq. 5, using different values for view factors and a bulk temperature following the standard fire. This boundary surface of the steel deck is subjected to standard fire conditions using a convection coefficient of 25 W/m^2K and an emissivity of fire equal to 1. The unexposed side is subjected to a convective heat flux (including the radiation heat flux), using a constant bulk temperature of 13 °C for slab 1 and 20 °C for slabs 2, 3 and 4. A convective coefficient of 9 W/m^2K is applied on this boundary surface of the composite slab in order to include the radiation effect [17].

With respect to the convergence criterion, the heat flow criterion is applied in ANSYS, using a tolerance value of 0.001 and a minimum reference value of 10^{-6}. In MATLAB, a value of 10^{-2} is set to the absolute tolerance and 10^{-1} to the relative tolerance, using a maximum number of Gauss-Newton iterations of 15.

5.2 Results

Figures 8, 9, 10 and 11 present the temperature development (numerical and experimental) at different selected points, as well as the average and maximum temperatures on the unexposed surface for each slab. The curves "ANS x_Py" and "MAT x_Py" refer to the results from ANSYS and MATLAB, respectively; where x is the thickness of the air gap and y is the number of the point. The curves "AVE_T" and "MAX_T" refer to the average and maximum temperature, respectively, on the unexposed surface of the slab. "EXPT" represents the experimental results.

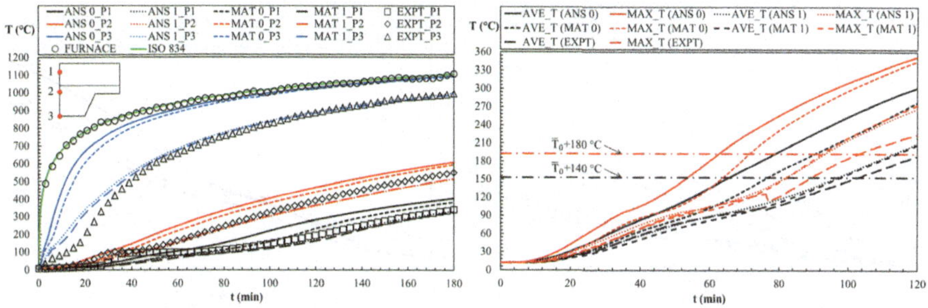

(a) Selected points. (b) Unexposed side.

Fig. 8. Numerical and experimental results for slab 1 - Points 1, 2 and 3 at distance 20, 70 and 130 mm from the top.

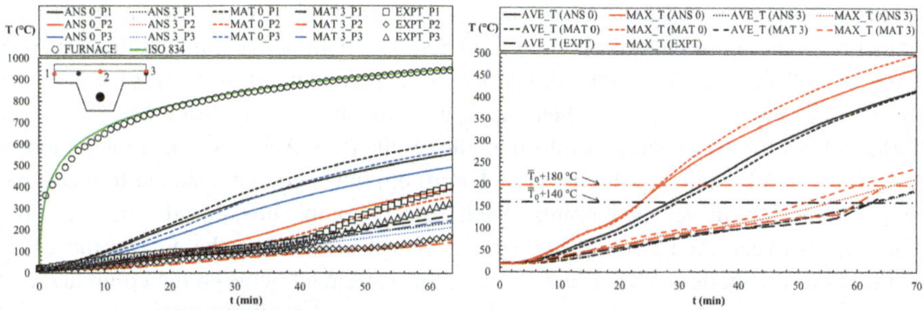

(a)Selected points. (b) Unexposed side.

Fig. 9. Numerical and experimental results for slab 2 - Points 1, 2 and 3 at distance 20, 15 and 15 mm from the top.

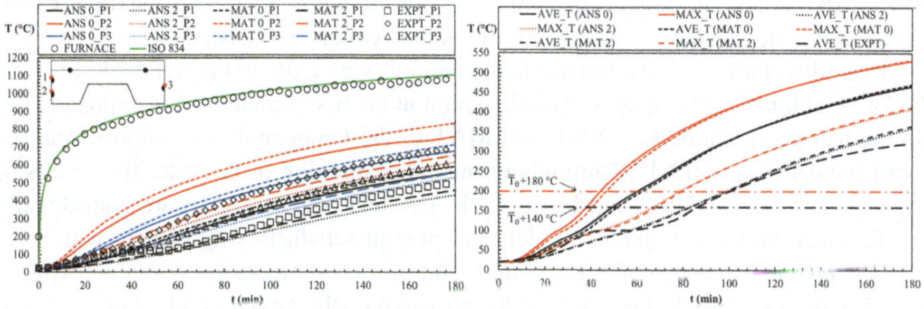

(a) Selected points. (b) Unexposed side.

Fig. 10. Numerical and experimental results for slab 3 - Points 1, 2 and 3 at distance 50, 85 and 70 mm from the top.

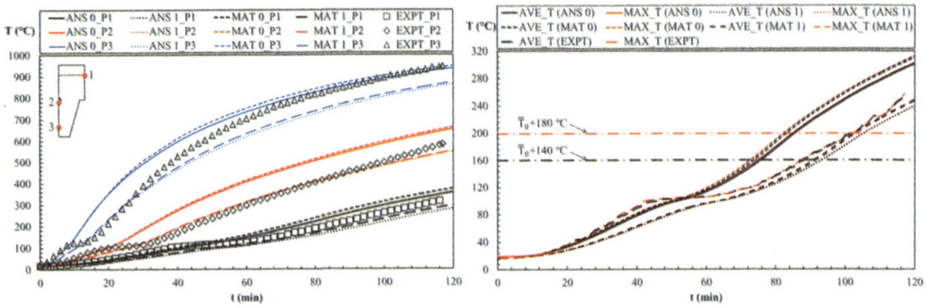

(a) Selected points. (b) Unexposed side.

Fig. 11. Numerical and experimental results for slab 4 - Points 1, 2 and 3 at distance 20, 74 and 123 mm from the top.

From the results of slab 1, a reasonable agreement is observed between the furnace temperature and the standard fire curve ISO 834. For both perfect contact models ANS 0 and MAT 0, a considerable difference to experimental results (EXPT) can be observed in the first stages of heating. The numerical results from MATLAB are slightly closer to experimental results than the results from ANSYS. The air gap models ANS 1 and MAT 1 (air gap thickness of 1 mm) provide a good agreement to measured temperatures for all selected points. With respect to the unexposed surface, a big difference between the results of the perfect contact models and the air gap models is verified. For the models ANS 1 and MAT 1, the agreement between the computed and measured temperatures is very good for both average and maximum temperatures until 76 min. After that, some differences are noticed, probably due to experimental deviations such as the furnace temperature, for example. In general, the results of the air gap models ANS 1 and MAT 1 are closer to each other in comparison to the perfect contact models ANS 0 and MAT 0. Small differences between the results of ANSYS and MATLAB models are observed, with exception of the average and maximum temperatures on the unexposed surface for the perfect contact models ANS 0 and MAT 0.

Regarding the slab 2, the furnace temperature is very close to the standard fire curve ISO 834, although presenting a small deviation in the first minutes of fire exposure. For the perfect contact models ANS 0 and MAT 0, the temperatures are higher than the experimental results (EXPT) throughout the entire duration of the test. The results of the air gap models ANS 3 and MAT 3 (air gap thickness of 3 mm) are considerably smaller than the perfect contact models and present satisfactory agreement with measured temperatures, principally for point 2 and average temperature on the unexposed side. For points 1 and 3, the results of the air gap models present good agreement with measured temperatures until the first 42 min of fire. Some differences are observed between the results of ANSYS and MATLAB models, being greater for the perfect contact models (ANS 0 and MAT 0).

For slab 3, it is noteworthy that the furnace temperature is under the standard fire curve ISO 834, which was used for the numerical simulations. Consequently, the numerical results should be higher than experimental results. For the perfect contact models ANS 0 and MAT 0, a good agreement with experimental results (EXPT) is observed for point 1 until the first 44 min of fire. After that, better results are obtained using the air gap models ANS 2 and MAT 2 (air gap thickness of 2 mm). This is because during a fire, the debonding of the steel deck from concrete occurs after a period of time. A delay in the rate of temperature increase is observed in the last minutes of fire on curve ANS 0_P2. It can be justified by the coincidence of two different materials (steel and concrete) on the same node and the abrupt change in the specific heat of steel for temperatures between 700 °C and 800 °C, see Fig. 7(a). In general, the results of the air gap models present a better agreement with measured temperatures in comparison to the results of the perfect contact models. For the selected points, models ANS 0 and MAT 2 present better agreement with experimental results when compared to models MAT 0 and ANS 2, respectively. Considerable differences between ANSYS and MATLAB models are evident, mainly for curves ANS 0_P2 and MAT 0_P2. On the other hand, the results of these models are very close to each other on the unexposed surface.

For slab 4, it can be observed that the temperature development on the selected points is quite similar between the experimental results EXPT and the perfect contact models ANS 0 and MAT 0 at the first minutes of heating. After that, better results are obtained using the air gap models ANS 1 and MAT 1 (air gap thickness of 1 mm), with exception to point 3. Regarding the temperature development on point 2, the air gap models present good agreement with experimental results for temperatures over 100 °C. For point 3, at the last minutes of heating, the perfect contact model presents better agreement with measured temperatures in comparison to the air gap model. Concerning the temperature development on the unexposed surface, the maximum and average temperature curves are very close to each other for all the models. In this case, better agreement with the experimental results is observed using the air gap models.

Tables 2, 3, 4 and 5 present the results obtained for the fire resistance according to the insulation criterion with respect to the average temperature rise (t_{fi} Ave) or the maximum temperature rise (t_{fi} Max) on the unexposed surface for each slab.

Table 2. Fire resistance of slab 1 (criterion I): experimental, numerical and analytical results.

	Model ANS 0	Model MAT 0	Model ANS 1	Model MAT 1	EXPT result	EN 1994-1-2
t_{fi} Ave (min)	65.6	76.7	97.0	97.7	*102.7*	*95.8*
t_{fi} Max (min)	*62.6*	*71.4*	*93.3*	*92.5*	103.0	

Table 3. Fire resistance of slab 2 (criterion I): experimental, numerical and analytical results.

	Model ANS 0	Model MAT 0	Model ANS 3	Model MAT 3	EXPT result	EN 1994-1-2
t_{fi} Ave (min)	28.3	29.7	*62.8*	62.3	*62.2*	*38.0*
t_{fi} Max (min)	*27.0*	*26.5*	65.9	*60.1*	–	

Table 4. Fire resistance of slab 3 (criterion I): experimental, numerical and analytical results.

	Model ANS 0	Model MAT 0	Model ANS 2	Model MAT 2	EXPT result	EN 1994-1-2
t_{fi} Ave (min)	49.4	51.4	82.6	82.9	*80.0*	*73.9*
t_{fi} Max (min)	*44.1*	*46.5*	*78.3*	*79.1*	–	

Table 5. Fire resistance of slab 4 (criterion I): experimental, numerical and analytical results.

	Model ANS 0	Model MAT 0	Model ANS 2	Model MAT 2	EXPT result	EN 1994-1-2
t_{fi} Ave (min)	*75.6*	*73.6*	*93.5*	*91.6*	*88.2*	*106.5*
t_{fi} Max (min)	84.8	82.0	105.1	102.5	102.1	

The results obtained with the air gap models underestimate the fire resistance for slab 1, with a relative error of 9.2% for ANSYS and 9.9% for MATLAB. Better approximation to experimental results is observed for the EN 1994-1-2 provisions, with a relative error of 6.7%.

With respect to the slab 2, a good agreement between the fire resistance obtained with the air gap models and experimental results is achieved, resulting in a relative error of 1.0% for ANSYS and 3.4% for MATLAB. An underestimated value is obtained using the EN 1994-1-2 provisions, with a relative error of 38.9%.

Regarding the slab 3, the results evidence that the fire resistance obtained using the air gap models is slightly underestimated, with a relative error of 2.1% for ANSYS and 1.1% for MATLAB. In addition, a good agreement between EN 1994-1-2 calculations and experimental data is also observed, resulting in a relative error of 7.6%.

According to the results for slab 4, it can be concluded that the air gap models slightly overestimated the fire resistance, with a relative error of 6.0% for ANSYS and 3.9% for MATLAB. A bigger discrepancy is obtained using the perfect contact models, with a relative error of 14.3% for ANSYS and 16.6% for MATLAB. The EN 1994-1-2 provisions overestimated the fire resistance, providing an unsafe result with a relative error of 20.7%.

In all experimental fire tests, the air layer which appears due to debonding of the steel deck from concrete does not have constant thickness. This means that the air gap is 0 mm thick at the beginning of the test (perfect contact) and as the specimen is heated, the steel deck separates slowly from the concrete, hence increasing the air gap thickness. By means of simplification, this work assumes that the air gap has constant thickness throughout the whole duration of the simulations.

6 Parametric Study

A parametric study has been conducted in order to evaluate the influence of the thickness of the concrete topping (h_1) and the air gap (t_a) on the fire resistance (I) of composite slabs. A total of 40 numerical simulations have been carried out, of which 32 performed in ANSYS, using the perfect contact model and the air gap model; and 8 performed in MATLAB, using the air gap model. In ANSYS, the air layer is divided into elements with 0.5 mm through the thickness.

The "O Feliz H60" trapezoidal steel deck profile (slab 2) has been selected to perform the numerical analyses. A representative portion of 1 m by 1 m of each slab has been considered to perform the thermal analyses. Table 6 presents the ranges of investigated parameters for the numerical simulations performed in ANSYS and MATLAB.

Table 6. Investigated parameters of the parametric study.

	h_1 (mm)	t_a (mm)
ANSYS	50, 60, 70, 80, 90, 100, 110, 120	0 (perfect thermal contact), 1, 2, 3
MATLAB	50, 60, 70, 80, 90, 100, 110, 120	3

Figure 12 presents the results for the fire resistance (I) obtained through the thermal models (ANSYS and MATLAB); experimental fire tests (EXPT); and simplified calculation method presented in Sect. 3. The experimental results were published in 1983 by the ECCS [3].

A model with quadratic dependence between the fire resistance t_i and the effective thickness h_{eff} is chosen to fit the results of the numerical simulations. A linear relationship between the effective thickness and the thickness of the air gap t_a is considered in order to consider the increase in fire resistance. The proposed new equation for the fire resistance of composite slabs is given in Eq. 8.

$$t_i = 0.0059 \cdot h_{eff}^2 + 0.1127 \cdot h_{eff} - 5.8065 + (0.1424 \cdot h_{eff} + 2.4672) \cdot t_a \qquad (8)$$

Where t_i is given in minutes, and h_{eff} and t_a are given in millimetres; with the following limits: $70 \leq h_{eff} \leq 150$ mm and $0 \leq t_a \leq 3$ mm.

Figure 12 illustrates the results obtained with the application of Eq. 8 (New Eq) for different air gap thicknesses.

Analysing the Fig. 12, a good agreement is observed between the results of the proposed new equation and the numerical results. A satisfactory agreement between the results of the EN 1994-1-2 and the numerical simulations is obtained using an air gap thickness of 2 mm. Good agreement between experimental and numerical results is observed considering an air gap thickness of 3 mm.

The thickness of the air gap t_a depends on the geometry of the profiled steel deck among other parameters. Further numerical analyses should be performed considering different geometries to define an optimal air gap thickness for composite slabs with several steel deck profiles.

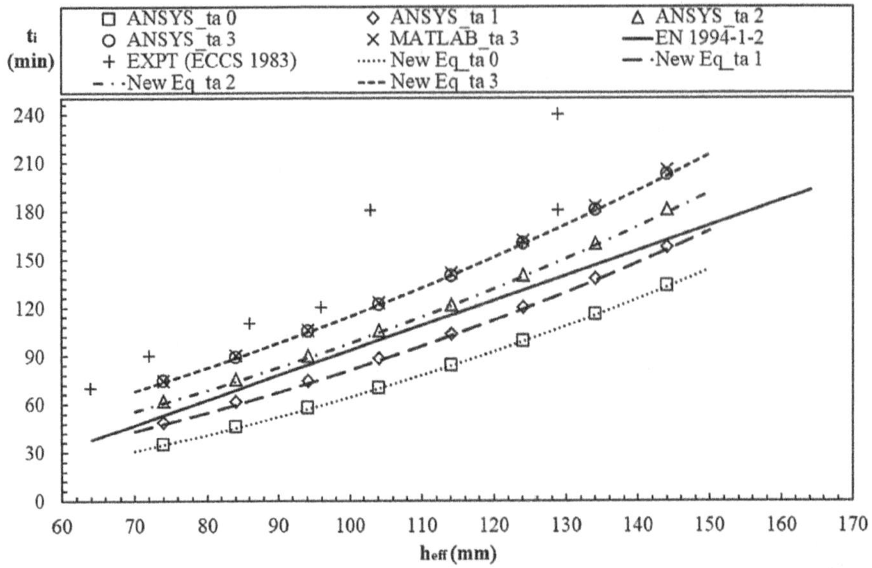

Fig. 12. Results of the proposed new equation for the fire resistance (I) of composite slabs.

7 Conclusions

This study presented a discussion around the results of three-dimensional thermal analyses performed in ANSYS and MATLAB for different composite slabs. The fire resistance according to thermal insulation criterion (I) was evaluated and compared to experimental results and simple calculation method of Eurocode 4 as well. With the aim of simulating the effects of debonding of the steel deck from concrete, an alternative thermal model was used, in which an insulating layer (air gap) with constant thickness was introduced between the concrete topping and steel deck.

For the experimental results of the four slabs, a plateau at about 100 °C (due to moisture evaporation) should be highlighted, consisting of a decrease in the rate of temperature increase. The results of the numerical simulations do not present this pronounced plateau, probably because localized moisture concentrations in the tests were higher than the uniform moisture content introduced in the thermal models for each slab.

The maximum temperature rise criterion was decisive to determine the fire resistance according to thermal insulation criterion for most of the simulations. With respect to the experimental results, the average temperature rise criterion governed the fire resistance. The Eurocode 4 provisions underestimated the fire resistance for all the slabs, and for slab 2 in particular, a considerable difference was observed. The perfect contact models underestimate the fire resistance. Therefore, it is evident that the air gap models provide much better results for fire resistance from the thermal insulation standpoint when compared to the perfect contact models, reducing the temperature rise on the selected points and unexposed surface as well. However, in order to determine the thickness of the air gap t_a, it is necessary to resort to experimental data. In this regard, further thermal analyses should be conducted to better understand the effect of the air gap on different slab profiles.

In some cases, the results of ANSYS and MATLAB models presented noticeable differences for the same situation, that is, for a same point and type of model (perfect contact or air gap model). These differences can be justified by the presence of the steel components within the concrete layer (rebar and mesh) in the ANSYS model which are not included in the MATLAB model. Although not significantly affecting the fire resistance (I), these components can affect the temperature development on particular points.

Based on the numerical results, a new formula has been proposed for the calculation of the fire resistance (I), depending on the effective thickness of the composite slab and the air gap thickness. This proposed model presents good agreement with numerical results and considers parameters which are not included in the current calculation rules of standards.

References

1. International Standard ISO 834: Fire-resistance tests - Elements of building construction (1999)
2. Jiang, J., Pintar, A., Weigand, J.M., Main, J.A., Sadek, F.: Improved calculation method for insulation-based fire resistance of composite slabs. Fire Saf. J. **105**, 144–153 (2019)
3. European Convention for Constructional Steelwork - Committee T3 - Fire Safety of Steel Structures, Calculation of the fire resistance of composite concrete slabs with profiled steel sheet exposed to the standard fire, Brussels (1983)
4. Hamerlinck, R.: The behaviour of fire-exposed composite steel/concrete slabs. Eindhoven University of Technology (1991)
5. Both, C.: The fire resistance of composite steel-concrete slabs. Delft University of Technology (1998)
6. Abdel-Halim, M.A.H., Hakmi, M.R., O'Leary, D.C.: Fire resistance of composite floor slabs using a model fire test facility. Eng. Struct. **21**(2), 176–182 (1999)
7. Lim, L., Wade, C.: Experimental fire tests of two-way concrete slabs. Christchurch (2002)
8. CEN - European Committee for Standardization, EN 1994-1-2: Design of composite steel and concrete structures - Part 1-2: General rules - Structural fire design. Brussels (2005)
9. CEN- European Committee for Standardization, EN 1994-1-1: Design of composite steel and concrete structures - Part 1-1: General rules and rules for buildings. CEN - European Committee for Standardization, Brussels (2004)
10. Guo, S., Bailey, C.G.: Experimental behaviour of composite slabs during the heating and cooling fire stages. Eng. Struct. **33**, 563–571 (2011)
11. Piloto, P.A.G., Prates, L.M.S., Balsa, C., Rigobello, R.: Numerical simulation of the fire resistance of composite slabs with steel deck. Int. J. Eng. Technol. (UAE) **7**(2.23), 83–86 (2018). (2018/04/206), ISSN 2227-524X. https://doi.org/10.14419/ijet.v7i2.23.11889
12. CEN - European Committee for Standardization, EN 1363-1: Fire resistance tests - Part 1: General Requirements. Brussels (2012)
13. CEN - European Committee for Standardization, EN 1365-2: Fire resistance tests for load bearing elements - Part 2: Floors and roofs (Withdrawal). Brussels (2014)

14. CEN - European Committee for Standardization, EN 1993-1-2: Design of steel structures - Part 1-2: General rules - Structural fire design. Brussels (2005)
15. CEN - European Committee for Standardization, EN 1992-1-2: Design of concrete structures - Part 1-2: General rules - Structural fire design. Brussels (2004)
16. Çengel, Y.A., Ghajar, A.J.: Heat and Mass Transfer: Fundamentals & Applications, 5th edn. McGraw-Hill Education, New York (2015)
17. CEN - European Committee for Standardization, EN 1991-1-2: Actions on structures - Part 1-2: General actions - Actions on structures exposed to fire. Brussels (2002)

Durability of Reaction to Fire Performance of Wood Based Panels Through Accelerated Aging Cycles

Luís Mesquita[1(✉)], Lucas Ferle[2], and Gerson Santos[3]

[1] Institute for Sustainability and Innovation in Structural Engineering (ISISE), Instituto Politécnico de Bragança, Campus de Santa Apolónia, 5301-857 Bragança, Portugal
lmesquita@ipb.pt

[2] Instituto Politécnico de Bragança, Campus de Santa Apolónia, 5301-857 Bragança, Portugal
lucascferle@gmail.com

[3] Departamento de Engenharia Mecânica, Universidade Tecnológica Federal do Paraná (UTFPR), Curitiba, Brazil
gsantos@utfpr.edu.br

Abstract. Wood and wood based products application in the construction is growing due to the increasing trend of sustainable development. Because of the EU policy about constructions, each time requiring improved products against fire hazard, there is a need of developing fireproof products to wood and wood based panels and know their life time behavior.

To evaluate the performance and durability of fire retarded wood based panels, concerning mechanical and reaction to fire behavior, an experimental study is done considering the long term behavior of wood based panels with and without fire retardant products after being submitted to accelerated aging and compared to non-aged wood based panels. Fire reaction was carried out through cone calorimeter tests to evaluate if the fire reaction properties were maintained or altered after exposure to environmental conditions.

In terms of mechanical properties, MDF and PB panels had significant losses in MOR and MOE values after being submitted to aging. This behavior is not so clear for OSB panels. Reduction in mechanical properties are smaller for MDF panels with flame retardant when compared to MDF without flame retardant whereas PB without flame retardant showed small losses compared to PB with flame retardant. MDF panels without flame retardant presented a decrease in the fire reaction properties, releasing more heat after aging, whereas in the PB and OSB without flame retardant properties were maintained. MDF and PB fire retardant panels reaction to fire behavior before and after aging allows to maintain them in class B of fire reaction.

Keywords: Durability · Accelerated aging · Wood based panels · Fire retardants · Fire reaction

P. A. G. Piloto et al. (Eds.): CILASCI 2019, LNCE 1, pp. 31–47, 2020.
https://doi.org/10.1007/978-3-030-36240-9_3

1 Introduction

Wood is being used by humans since the early civilizations, and was one of the most important materials used the building construction. In recent years, due to ecological and environmental policies and restrictions in Europe, wood, wood products and wood structural elements have being positioned as a green raw material, Ecologically Sustainable and renewable material with a positive impact in the buildings carbon dioxide emissions in comparison to other construction materials, such as steel, concrete and bricks. The EUs driving policies for a competitive economy with low carbon emissions, [1], boost its architectural and engineering application in the building industry, but actually subjected to an higher demand in terms of its life cycle performance basic requirements, such as the sustainable use of natural resources, mechanical resistance and stability and Safety in case of fire, among others, [2].

The disseminated use of wood and wood products in the building construction have led to a need of wood based product development (Engineered wood products), namely wood-based panels, such as particle board (PB), medium density fibreboard (MDF), plywood, hardboards and wood flooring, [3], and wood structural members from large wood panel construction using cross-laminated timber (CLT), and others [4].

Being a hygroscopic material, wood thermal and mechanical properties, and aesthetic appearance, are affected by its surrounding environment, regarding temperature, humidity and direct or indirect solar radiation in outdoor and indoor appliances. Furthermore, when the moisture content is above 20%, wood is susceptible to attack by fungi and bacteria. Structural wood products when exposed to excessive moisture variations can lead to swelling or shrinkage causing warping and cracking of the element reducing its mechanical properties, stability and durability. Additionally, wood structural elements with superficial cracks will have their reaction and resistance to fire reduced as the fire will propagate through them leading to a faster cross section charring rate and heat release rate (HRR). For these reasons different wood treatment methods, physical or chemical treatments, are used to increase wood stability and durability, and improving the resistance to biological degradation, fire resistance, UV resistance and mechanical properties, [4, 5]. Currently applied superficial chemical treatments include coating moisture-, bio-, fire- or UV-resistant agents on the surface of wood.

Wood is considered a flammable material, and although it has an intrinsic/natural fire protection, charring to decrease the heating rate, from the European standard fire classification of construction products and building elements, EN13501-1 [6], untreated wood is usually classified as being of class D, with lower density products in class E. This classification system considers the reaction to fire performance, smoke production and flaming droplets/particles. When fire retardant treatments are applied wood products can reach C and B class levels. Table 1 shows how the classification of construction products is made based on fire reaction levels [7].

Table 1. Classification of the reaction to fire of wood products.

Euro class	Smoke class	Burning droplets class	Typical products
A1	–	–	Stone, concrete
A2	s1 s2, or s3	d0 d1 or d2	Gypsum boards (thin paper), mineral wool
B	s1 s2, or s3	d0 d1 or d2	Gypsum boards (thick paper), fire retardant wood products
C	s1 s2, or s3	d0 d1 or d2	Coverings on gypsum boards
D	s1 s2, or s3	d0 d1 or d2	Wood, wood-based panels
E	–	– or d2	Some synthetic polymers
F	–	–	No tested or incapable of achieving Class E

Additionally, when wood products are protected with non-fire retardant coatings their ignition properties and flame spread are influenced by the coating chemical composition and film thickness, [8, 9]. Wood treatment with fire retardant coatings (FRC) or intumescent fire retardant coatings (IFRC), [10], can overcome these weaknesses when wood products are exposed to fire and, for wood structural elements, assure the required fire resistance and load bearing capacity to be used in the building construction, meeting the requirements of the Eurocode 5, [11], based on the National fire safety codes requirements.

Table 2. Requirements for DRF Classes of fire-retardant wood products in interior and exterior end use applications, [12].

DRF class		Existing fire requirements	Additional performance requirements at different end use of fire retardant wood-based products	
	Intended use	Reaction to fire class, initial	Hygroscopic properties	Reaction to fire performance after weather exposure
INT1	Interior dry applications	Relevant fire class	–	–
INT2	Interior humid applications	Relevant fire class	– Moisture content <28% – No exudation of liquid – Minimum visible salt with no increase at surface	–
EXT	Exterior applications	Relevant fire class	– Moisture content <28% – No exudation of liquid – Minimum visible salt with no increase at surface	Maintained reaction to fire performance -after – Accelerated weathering or – Natural weathering Application of specified maintenance may be included

Fire retardants applied in the products surface or by pressure impregnation may considerably improve the fire properties of wood and wood products, but the long term durability of this protection is not fully known. It is expected that, mainly in exterior applications but also in interior humid conditions, the fire retardant efficiency may reduce due to its hygroscopicity, [13, 14] and water solubility of the chemicals used.

The recent standard EN16755, [12], specifies a new classification testing for Durability of Reaction to Fire performance (DRF) based mainly on the Nordtest standard NT Fire 054 [15], which was based on earlier ASTM standards, namely the D2898-94. This classification is based on the intended use, considering interior dry and humid applications and exterior applications, as shown in Table 2. For exterior applications, the reaction to fire performance after weather exposure can be classified using natural or accelerated weathering.

To evaluate the performance and durability of fire treated wood based panels on the thermal and mechanical properties, including reaction to fire, a study is being done considering the long term behavior of wood products with and without fire retardant products after being submitted to accelerated aging and compared to non-aged wood products.

A set of experimental tests are performed towards the mechanical characterization and fire reaction of different wood based panels with and without fire retardant products, according to the EN 310 standard [16] to determine bending strength (MOR) and modulus of elasticity (MOE), and in the cone calorimeter to evaluate mass loss and heat release rate.

2 Materials and Methods

2.1 Wood Based Panels Artificial Aging

The wood panels were aged artificially using a cycle of humidity and temperature for indoor environments. Although ETAG 028:2012, [9], is specifically applied to construction products protected by paints, varnishes and surface-impregnated products, and for products with FR treatments applied on construction sites, it was used as a reference for the definition of the aging cycle, depending on the product category of use. The categories of use are referred as type X, used in internal and external applications and exposed to rain and ultraviolet radiation, type Y, intended for indoor and outdoor environments not exposed to rain or UV, Z1 used in indoor environments exposed to high humidity and Z2 used for internal use only.

The cycle used in this paper reproduces the category of use of type Z1, similar to DRF class INT 2 in EN 16755, in which the panels are exposed to 27 ± 2 °C and 90 ± 5% relative humidity during 8 h, following of 16 h at 23 ± 2 °C and 50 ± 3% relative humidity, resulting in a 24-hour cycle, carried out for 10 days without interruption for an expected life of 5 years, [10]. Figure 1 shows the experimental sequence for the mechanical characterization and reaction to fire analysis of aged and non-aged wood based samples.

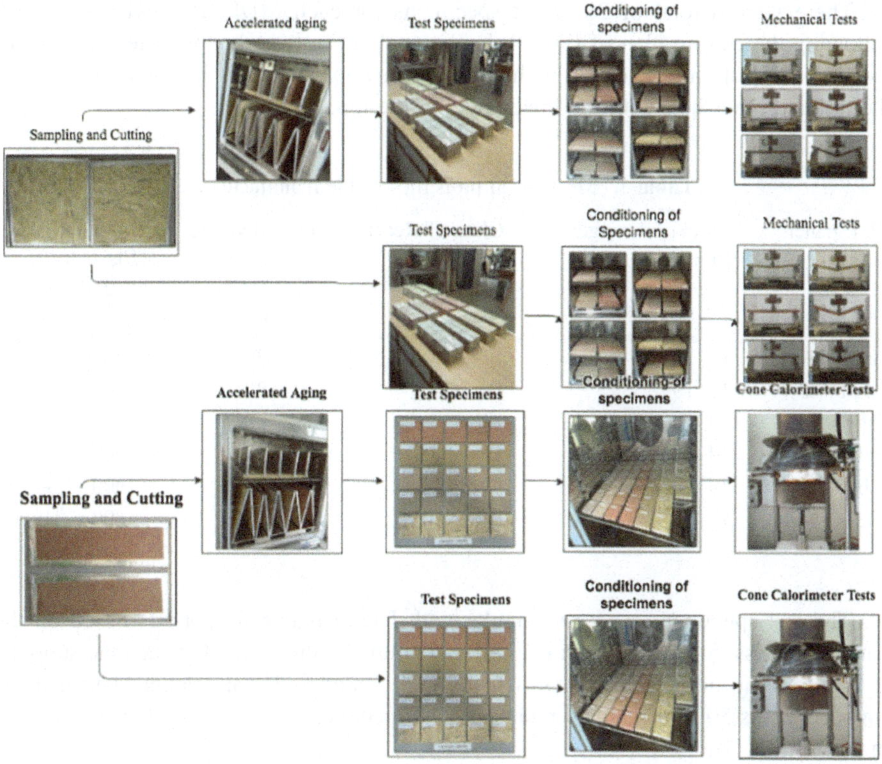

Fig. 1. Sequential procedure used for mechanical and reaction to fire tests.

2.2 Mechanical Characterization

The mechanical characterization was done in five different wood based panels, with and without fire retardant: standard Medium Density Fiberboard without fire retardant (MDF-ST-NFR), Medium Density Fiberboard with fire retardant (MDF-FR), Particle Board type P2 without fire retardant (PB-P2-NFR), Particle Board type P2 with fire retardant (PB-P2-FR) and Oriented Strand Board type 4 without fire retardant (OSB3-NFR). The panels were all supplied by the company Sonae Arauco, [17]. The FR components are mixed and added to de resin during the panels production process. The type of resin used in the production varies. The UF/MUF, Urea-Formaldehyde and Melamine-Urea-Formaldehyde, resin is used in MDF and PB panels, while pMDI, polymeric diphenyl methane diisocyanate, resin is used in the OSB production.

The mechanical properties of the panels provided by the manufacturer and its fire reaction classes are shown in Table 3.

The mechanical strength was determined by the standard EN310 [16], using the three-point bending test to determine bending strength (MOR) and Modulus of elasticity (MOE).

There were a total of 20 tested specimens for each MDF, PB and OSB panel, following the cutting plan of EN310. Each panel cut in two groups of ten specimens, for each orientation 0° and 90°, with half of the samples tested with the upper side on the top and other half with the lower side on the top.

Table 3. Mechanical properties of the manufacturer.

Ref. panel	Thickness ranges [mm]	Class of Reaction to fire	Bending strength [MPa]		Modulus of Elasticity [MPa]	
			0°	90°	0°	90°
MDF-FR	13–19	B-s2, d0	20	–	2200	–
MDF-ST-NFR	13–19	D-s2, d0	20	–	2200	–
PB-P2-FR	14–20	B-s1, d0	11	–	1600	–
PB-P2-NFR	14–20	D-s2, d0	11	–	1600	–
OSB3-NFR	18–25	D-s2, d0	26	14	4800	1900

The test specimens were rectangular with length between supports based on the panel thickness. Since the width is b (50 ± 1) mm and the length between the supports is 20 times the nominal thickness (t), the total length is l1 mm (length between the supports), plus 50 mm. Table 4 represents the specimens size for each panel type used in the tests.

Table 4. Dimensions of test pieces used in the test.

Types of panels	N° of test pieces	Width (b) [mm]	Thickness (t) [mm]	Length between the supports (l1) [mm]	Total length (l2) [mm]
MDF-FR-0°	10	50	16	320	370
MDF-FR-90°	10	50	16	320	370
MDF-NFR-0°	10	50	16	320	370
MDF-NFR-90°	10	50	16	320	370
PB-FR-0°	10	50	15	300	350
PB-FR-90°	10	50	15	300	350
PB-NFR-0°	10	50	15	300	350
PB-NFR-90°	10	50	15	300	350
OSB3-NFR-0°	10	50	18	360	410
OSB3-NFR-90°	10	50	18	360	410

The test specimens were conditioned in a climatic chamber (ACS DM600) to a constant mass, for all the samples to enter in a hygroscopic equilibrium in an atmosphere with relative humidity of (65 ± 5)% and a temperature of (20 ± 2) °C, according to Fig. 2. Conditioning of test specimens. It was considered that a constant mass was reached when the results of two consecutive measurements of the test piece mass, carried out at 24 h of distance, are not differing of more than 0,1%, which means that the test piece mass cannot differ more than 0.10 g. Eight days of conditioning were necessary so that the constant mass be reached.

Fig. 2. Conditioning of test specimens.

The three point bending test was done using an Universal testing machine suitable for bending tests up to 100 [kN], INSTRON 3382. The setup consists of a cylindrical load head with 30 [mm] diameter placed parallel to the supports at the specimen mid span, as in Fig. 3. The supports are adjustable to allow the different length specimens support on a cylindrical clamp with 15 [mm] diameter, as shown in Fig. 3.

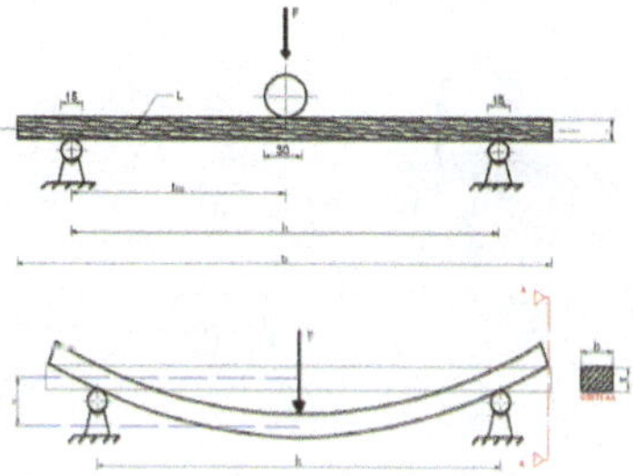

Fig. 3. Schematic representation of the test and measurement of deflection, [16].

The load was applied at a rate determined to achieve the maximum load within 60 ± 30 s throughout the test. The mid span vertical displacement was also measured during the tests.

The bending strength calculation (MOR) was calculated from the following Eq. 1.

$$MOR = \frac{3F_{máx}\, l_1}{2bt^2} \tag{1}$$

Where $F_{máx}$ represents the maximum load (N), l_1 is the distance between the centers of the two supports (mm), t is the thickness of the test specimens (mm) and b is the width of the test specimens.

For the modulus of elasticity (MOE) calculation, it was necessary to use Eq. 2, having a direct relationship between MOE and the maximum strength obtained in the bending test. The way in which the MOE should be calculated in the sample elastic regime, as proposed by the EN 310 standard [16], uses F_1 corresponding to 10% of the max break strength and F_2 corresponding to 40% of the α_1 and α_2 deformations.

$$MOE = \left[\frac{l_1^3(F_2 - F_1)}{4bt^3(\alpha_2 - \alpha_1)}\right] \tag{2}$$

2.3 Reaction to Fire Performance of Wood Based Panels

For the evaluation of the reaction to fire, a mass loss calorimeter was used according to ISO 5660, [13], see Fig. 4b. The samples were exposed in horizontal orientation to a radiant heat flux of 50 [kW/m^2] positioned at a distance of 25 [mm] from the cone base. An ignitor was positioned above the test sample since the beginning of the test and removed when the material ignition started. During the tests, the following parameters were obtained: heat release rate (HRR), ignition time (IT), total heat release (THR) and residual mass (m/m0). The dimensions used were 100 [mm] × 100 [mm] × thickness, indicated in Table 5. This table also shows the number of tests used with and without accelerated aging, referenced as AGED and NAGED, respectively, and the fire reaction class obtained from the product data sheet.

(a) (b) (c) (d)

Fig. 4. (a) Samples at the climate chamber. (b) Cone calorimeter test. (c) Samples before fire reaction test. (d) Samples after fire reaction test.

Table 5. Number of specimens tested in the cone calorimeter.

Panel type	Fire reaction class	N° of samples		Thickness [mm]
		NAGED	AGED	
MDF-FR	B-s2, d0	4	4	16
MDF-ST-NFR	D-s2, d0	4	5	16
PB-P2-FR	B-s1, d0	4	5	15
PB-P2-NFR	D-s2, d0	5	4	15
OSB3-NFR	D-s2, d0	4	5	18

Before the tests, the samples were also submitted to the atmospheric conditioning to reach the hygroscopic equilibrium at temperature of (23 ± 2) [°C] and relative humidity (RH) of (50 ± 5) [%].

The results of the heat release rate and the total heat released measured by the tests carried out in the calorimeter allows to evaluate the durability, through the accelerated aging cycles, on the fire reaction of the samples, and thus to reclassify the fire reaction of the samples. This performance is analyzed with reference to EN 16755 [12], which considers the criteria presented in Table 6 for the definition of the reaction to fire after climatic exposure.

Table 6. DRF classification criteria for small scale fire testing after weather exposure, [12].

	Building products excluding floorings
Heat flux	50 kW/m^2
Criteria for small scale fire testing after weather exposure	Class B products (according to EN 13501–1): Heat Release Rate, HRR30s ave $\leq$ 150 kW/m^2 during 600 s after ignition and Total Heat Release THR600s increase <20% compared to fire testing before the weather exposure Class C products (according to EN 13501–1): HRR 30s ave $\leq$ 220 kW/m^2 during 600 s after ignition and THR600s increase <20% compared to fire testing before the weather exposure

3 Results and Discussions

The most distinctive property of the MDF panels is its homogeneous composition, due to their reduced particles size. Thus, the mechanical properties between the test specimens do not vary much, regardless the orientation of the panel cut. An MDF panel feature is that outer layers have a higher density compared to inner layers, it follows that the outer layers have a higher compaction, occasionally causing greater mechanical resistance compared to other panel types, [18].

The experimental results from the MDF wood based panels are shown in the Figs. 5 and 6 for the fire retardant and non-fire retardant panels, respectively.

Considering the non-aged samples, the average values for the 0° orientation test specimens were of 30.21 [MPa] for MOR and 3.18 [GPa] for the MOE. For the 90° orientation those values are of 29.53 [MPa] and 3.21 [GPa]. For panels without fire retardant the mean values of MOR and MOE were 32.83 [MPa] and 3.13 [GPa] for the test specimens at 0°, and for the values at 90° the MOR and MOE was 31.87 and 3.12 [GPa].

Comparing the results from non-aged and aged tests, can be verified a reduction of MOR and MOE in both MDF board types with aging. The highest perceptual reduction is always for the 90° test results in both mechanical properties. Equally an higher reduction of both properties is obtained with aging for MDF-NFR panel, in comparison to the fire retardant panel (MDF-NFR).

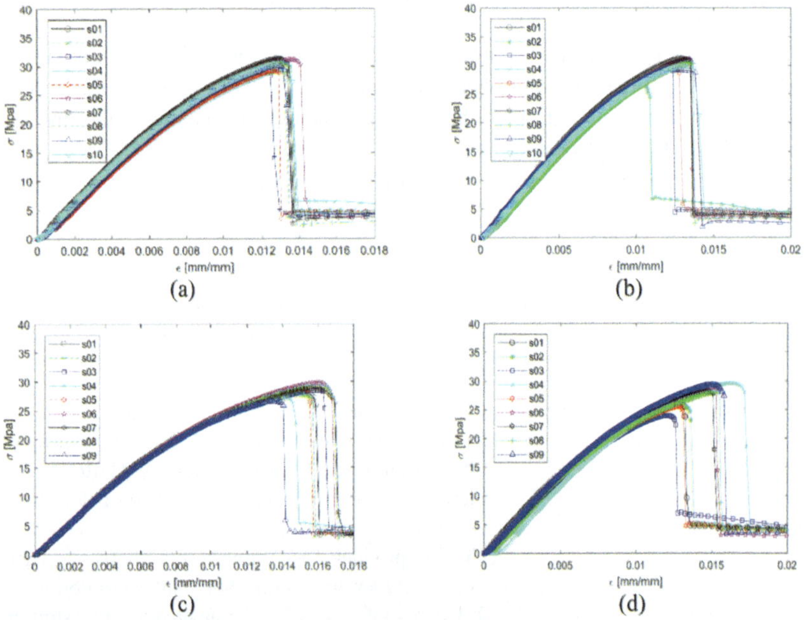

Fig. 5. Bending strength MDF-FR: (a) Direction 0° NAGED, (b) Direction 90° NAGED, (c) Direction 0° AGED, (d) Direction 90° AGED.

The particleboard panels have the most consistent values among those provided, due to the reduced size of its particles and high degree of homogeneity. The test results are represented in the next figures (Figs. 7 and 8).

There was apparently no significant variation of MOR and MOE in both directions, but fire-retardant panels had a higher modulus of elasticity and a small variation compared to MOR values. The mean values of MOR and MOE for PB-P2-FR were of 11.10 [MPa] and 2.54 [GPa] respectively for the 0° direction. The mean values for the specimens tested at 90° were of 11.85 [MPa] and 2.73 [GPa] respectively.

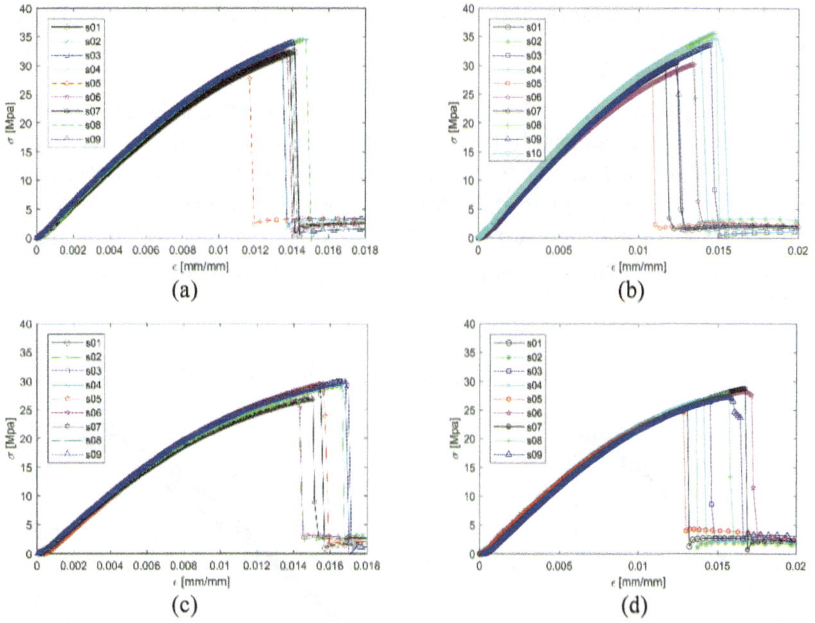

Fig. 6. Bending strength MDF-ST-NFR: (a) Direction 0º NAGED, (b) Direction 90º NAGED, (c) Direction 0º AGED, (d) Direction 90º AGED.

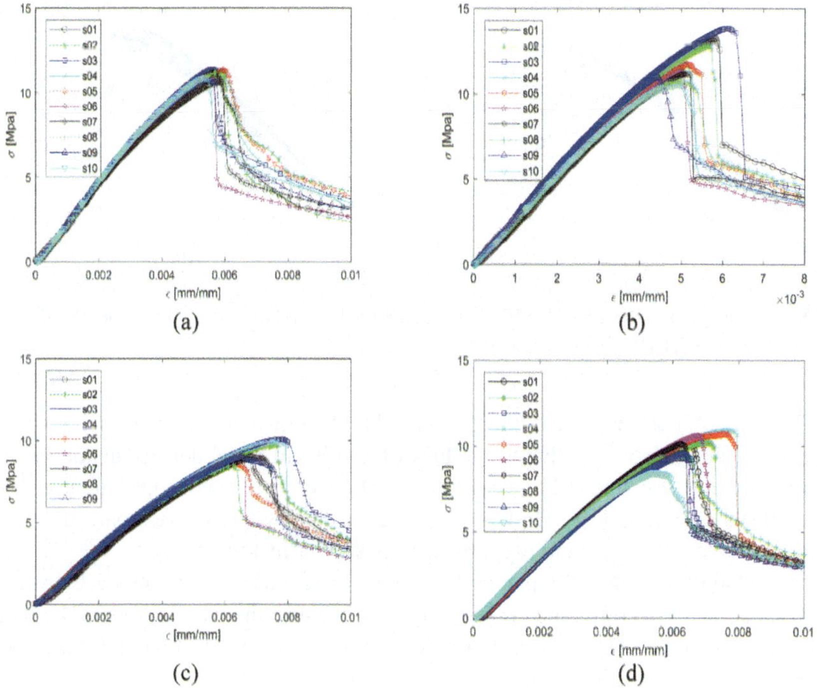

Fig. 7. Bending strength PB-P2-FR: (a) Direction 0º NAGED, (b) Direction 90º NAGED, (c) Direction 0º AGED, (d) Direction 90º AGED.

For the PB-P2-NFR panels, the mean values were of 11.59 [MPa] and 2.31 [GPa] at 0° for the MOR and MOE values, respectively, and for the 90° tests, were of 11.52 [MPa] and 2.35 [GPa].

The mechanical properties deterioration with the accelerated aging was more significate for this type of panels. In the case of PB-P2-FR the reduction was 16.84% for MOR and 31.38% for MOE at 0° direction, and for the 90° direction a reduction of 16.01% and 26.41% was obtained for MOR and MOE, respectively.

For the standard particleboard panel, PB-P2-NFR, a similar reduction magnitude was obtained after aging the samples, being the highest reduction also for the 0° direction tests.

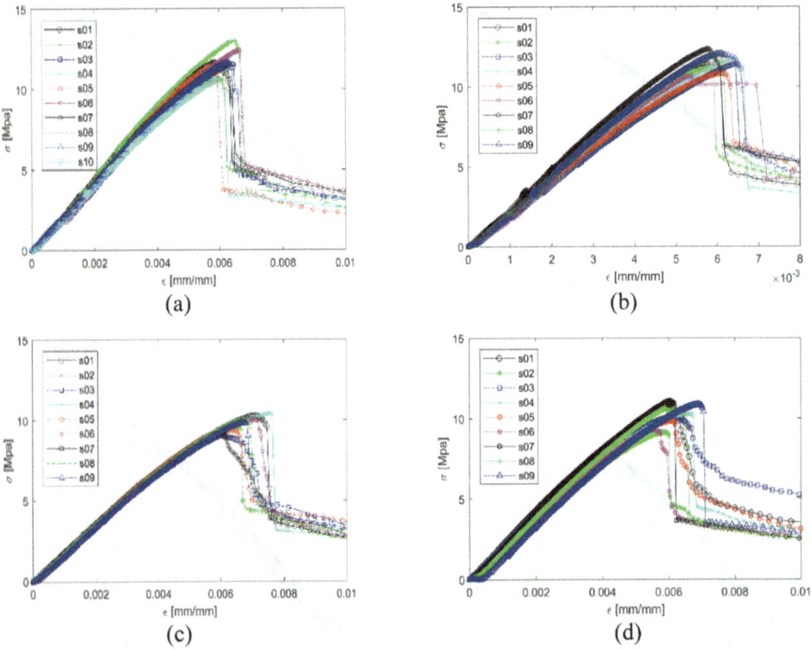

Fig. 8. Bending strength PB-P2-NFR: (a) Direction 0° NAGED, (b) Direction 90° NAGED, (c) Direction 0° AGED, (d) Direction 90° AGED.

The OSB panels presented more dissimilar flexural strength values between the specimens. This behavior is due to the lack of a uniform panel density inside the plate, this implies that specimens have a higher surface density, and consequently, higher values of static bending [19]. However, the higher density in the lower part of the board implies smaller values of bending strength, as shown in Fig. 9.

The behavior of the OSB panels has shown a remarkable difference between the two orientations. This difference is so significant because the wood fibres in the parallel orientation are better organized and oriented to counter the pressure and therefore resist to higher values of tension.

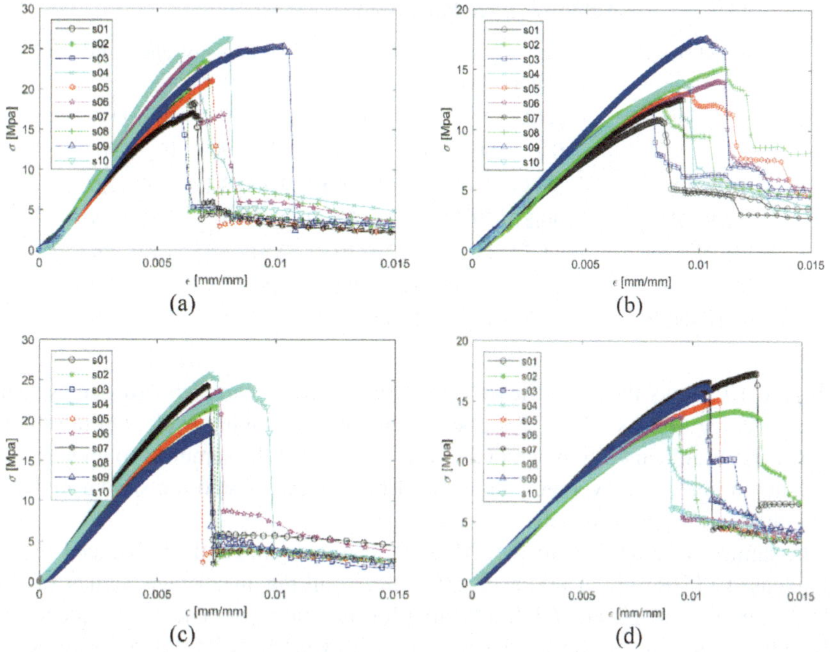

Fig. 9. Bending strength OSB-NFR: (a) Direction 0º NAGED, (b) Direction 90º NAGED, (c) Direction 0º AGED, (d) Direction 90º AGED.

A significant difference was observed in the test specimens values having the same orientation. The average value found for the panels tested at an orientation of 0° was 21.80 [MPa] for MOR and 3.93 [GPa] for MOE. In the panels tested at 90° the values of the analyzed mechanical properties are significantly reduced, resulting in MOR and MOE values of 13.56 [MPa] and 1.88 [GPa], respectively.

OSB panel samples present the smallest variation with the aging effect. In fact considering the arithmetic average of all samples, the MOR had a small increase in both directions, while the highest MOE reduction was obtained at 0º direction, with a value of 5.75%. Considering the test dispersion of results, verified by the MOR standard deviation values, no definitive conclusion can be obtained about the aging effect on the OSB panels. A further experimental work is needed.

The complete experimental three point bending test results performed to all wood based panels are presented in Table 7. The table shows the average values of the Bending strength (MOR) and the Modulus of elasticity (MOE) for both directions (0º and 90º) of all samples before and after the accelerated aging.

Table 7. Three Point bending test results.

Wood based panel	Bending strength (MOR) [MPa]				Modulus of elasticity (MOE) [GPa]			
	NAGED		AGED		NAGED		AGED	
	0°	90°	0°	90°	0°	90°	0°	90°
MDF-NFR	32,83	31,87	28,61	26,87	3,13	3,12	2,67	2,60
MDF-FR	30,21	29,53	28,35	27,32	3,18	3,21	2,82	2,81
PB-P2-NFR	11,59	11,52	9,77	10,21	2,31	2,35	1,81	2,07
PB-P2-FR	11,10	11,85	9,23	9,95	2,54	2,73	1,75	2,01
OSB3-NFR	21,80	13,56	22,09	14,76	3,93	1,88	3,71	1,83

Figure 10 shows the heat release rate variation from particleboard panels, with a moving average of 30 s (HRR_30s), before and after the aging cycle. During the tests there was no significant difference in the residual mass of the samples, so their variation is not shown here. The average values of all the samples tested are shown in Table 7 for the different types of panels.

The aging cycle applied to the PB-NFR panels significantly influenced the total heat released (THR) up to 600 s of exposure, resulting in average values of 45.71 [MJ/m^2] for aged panels and 63.36 [MJ/m^2] for the non-aged. There was a decrease in THR between non aged and aged panels of 27.8% and 16.5% for products without flame retardant and flame retardant, respectively.

Exposure to temperature and humidity cycles causes changes in the behavior of urea-formaldehyde resin by releasing volatile compounds, reducing their contribution to the combustion of the panel. The ignition time of PB-NFR-NAGED was 32.2 [s] and the PB-NFR-AGED was 41.75 [s].

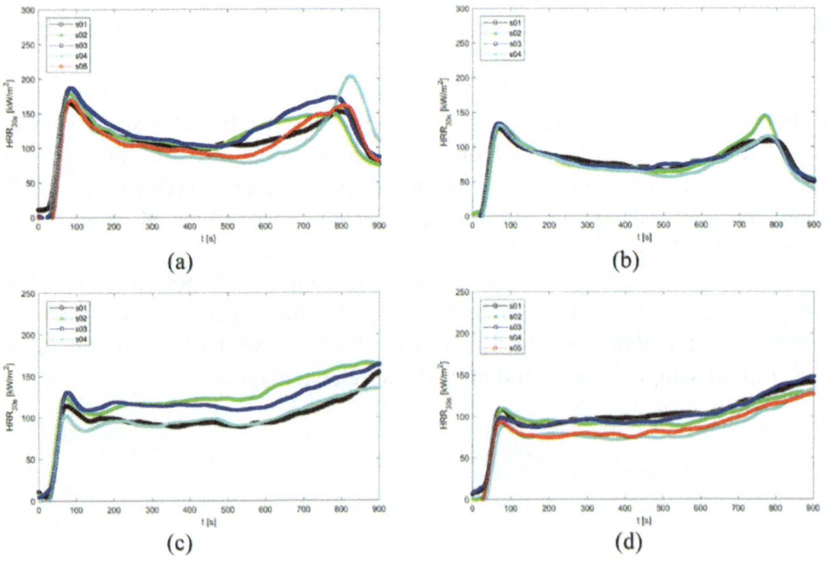

Fig. 10. HRR results. (a) PB-NFR and NAGED. (b) PB-NFR and AGED. (c) PB-FR and NAGED. (d) PB-FR and AGED.

With the exception of MDF panels without flame retardant, exposure to accelerated aging leads to a reduction in the rate of heat release as shown in Table 8. In the MDF-NFR samples the THR increased from 56.87 to 67.75 [MJ/m^2] after aging, while the ignition time decreases from 38.5 to 33.5 [s] in aged panels. In the case of MDF panels with fire retardant, there was no ignition and THR was lower in aged samples, similar to PB panels, changing from 11.35 to 11.03 [MJ/m^2].

Table 8. Results and comparison between aged and non-aged wood based panels.

Panel type	HRR_{30s} [kW/m^2]		THR_{600s} [MJ/m^2]		TI [s]	
	NAGED	AGED	NAGED	AGED	NAGED	AGED
PB-P2-FR	108.183	93.228	58.706	49.024	38.0	42.2
PB-P2-NFR	111.555	75.038	63.360	45.711	32.2	41.75
MDF-FR	28.456	21.902	11.356	11.033	–	–
MDF-ST-NFR	94.094	120.916	56.875	67.754	38.5	33.5
OSB4-NFR	184.401	122.820	103.83	65.377	32.0	29.4

The behavior observed in the OSB-NFR panels was similar to PB-NFR, in which the THR decrease after the aging of the samples, releasing less heat. The ignition time also decreased in relation to the non-aged samples, from 32 to 29.4 [s].

The reaction to fire performance after accelerated aging exposure of wood based panels with fire retardants (PB-P2-FR and MDF-FR), considering the rate of heat release and the total heat released, according to Table 6, allows to classify them in class B of reaction to fire after the exposure to accelerated aging.

4 Conclusions

Wood based panels is being used in building construction as a construction product. To overcome the lack of fire reaction it is frequent that wood based panel's producers to offer panels with fire retardants. It is not fully known how this panel behave in the long term, or if they are able to maintain their fire reaction classification when exposed to accelerated aging conditions (humidity and temperature variations). The main goal of this study is to give some clarification about the durability of fire reaction performance of wood based panels with and without fire retardants.

This work presented a set of experimental tests to determine mechanical properties of MDF, PB and OSB wood based panels. The Bending strength (MOR) and Modulus of elasticity (MOE) determined agree with the boards manufacture, except for the case of MDF panels where a difference of about 10 [MPa] was verified.

Also the MDF panels tests performed at 0° and 90° do not showed significant variation, due to panel homogeneity. However the behavior of OSB4 at 0° and at 90° is very different, presenting MOR and MOE values about 40% and 55% smaller, respectively.

With the exception of MDF-NFR panels, there was a decrease in the total heat released in the tests performed on the samples after exposure to accelerated aging. The aging of OSB panel without flame retardants resulted in a 37% decrease in THR compared to the non-aged. Considering the DRF classification criteria for small scale fire testing, specified by EN 16755 and mentioned in Table 6, the durability performance analysis allows to maintain the analyzed flame retardant panels in class B of fire reaction.

Acknowledgments. The authors acknowledge the support given by the company Sonae Arauco for providing the wood based panels.

References

1. European Commission, A Roadmap for moving to a competitive low carbon economy in 2050. European Commission: Brussels (2011)
2. JOUE, EU Construction Products Regulation (No. 305/2011) E. Parliament, Editor. Official Journal of the European Union (2011)
3. Lee, B.-H., et al.: Evaluating the flammability of wood-based panels and gypsum particleboard using a cone calorimeter. Constr. Build. Mater. **25**(7), 3044–3050 (2011)
4. Ramage, M.H., et al.: The wood from the trees: the use of timber in construction. Renew. Sustain. Energy Rev. **68, Part 1**, 333–359 (2017)
5. Esteves, B.M., Pereira, H.M.: Wood modification by heat treatment: a review. BioResources **4**(1), 370–404 (2009)
6. CEN, EN 13501-1: Fire classification of construction products and building elements: part 1 classification using test data from reaction to fire tests. European Committee for Standardization: Brussels, Belgium, p. 41 (2002)
7. Östman, B., Mikkola, E.: European classes for the reaction to fire performance of wood products. Holz als Roh- und Werkstoff **64**(4), 327–337 (2006)
8. Luo, J.: Ignition properties of panels coated with finishing fire-retardant paints under external radiation. Procedia Eng. **135**, 123–127 (2016)
9. Harada, T.: Time to ignition, heat release rate and fire endurance time of wood in cone calorimeter test. Fire Mater. **25**(4), 161–167 (2001)
10. Daniliuc, A., et al.: New trends in wood coatings and fire retardants: biobased monomers and high performance coatings. European Coatings Journal (2012)
11. CEN, EN1995-1-2 Eurocode 5: Design of timber structures - Part 1-2: General - Structural fire design. European Committee for Standardization: Brussels, Belgium (2004)
12. CEN, EN 16755: Durability of reaction to fire performance - Classes of fire-retardant treated wood products in interior and exterior end use applications. European Committee For Standardization: Brussels (2017)
13. Östman, B., et al.: Durability of fire retardant treated wood products at humid and exterior conditions review of literature. Fire Mater. **25**(3), 95–104 (2001)
14. Östman, B.A.L., Tsantaridis, L.D.: Durability of the reaction to fire performance of fire-retardant-treated wood products in exterior applications – a 10-year report. Int. Wood Prod. J. **8**(2), 94–100 (2017)
15. NORDTEST, NORDTEST NT Fire 054, Durability of reaction to fire – Performance classes of fire-retardant treated wood-based products in interior and exterior end use applications. Nordic Innovation Centre (2006)

16. CEN, EN310 Determination of modulus of elasticity in bending and of bending strength 1993, European Committee for Standardization
17. SonaeArauco, Construction and furniture products: technical data. www.sonaearauco.com
18. Torquato, L.: Caracterização dos painéis MDF comerciais produzidos no brasil, in Engenharia Florestal, Setor de Ciências Agrárias. Universidade Federal do Paraná (2008)
19. Del Menezzi, C.: Estabilização dimensional por meio do tratamento térmico e seus efeitos sobre as propriedades de painéis de partículas orientadas (OSB). Universidade Federal do Paraná (2013)

Modelling Real Fire by FDS and 2-Zone Model for Structural Post-Fire Assessment

Tom Molkens[1(✉)] and Barbara Rossi[1,2]

[1] Department of Civil Engineering, KU Leuven, Technology Campus de Nayer, Jan Pieter de Nayerlaan 5, 2860 Sint-Katelijne-Waver, Belgium
tom.molkens@kuleuven.be

[2] Department of Engineering Science, University of Oxford, Parks Road, Oxford OX1 3PJ, UK

Abstract. Post-fire resistance assessment of industrial structures is of prime importance to companies having to deal with such accidental situations. Most of the time, the structure or at least a major part of it still stands. Being able to quickly assess the temperature it was once submitted to, is very important to reevaluate its load-bearing capacity. The latter is to help take wise decisions regarding its dismantling or replacement and to avoid unnecessary delays during which the industry can no longer carry out its business. This paper describes a quick methodology to do so and demonstrates its accuracy with a case-study. Although full-scale fire experiments are reported in the literature, they are mostly under-instrumented, in such a way that few information is usually available. However, in 2018, a full-scale and fully instrumented fire experiment was conducted by Tongji university on a steel frame single-story building. This case study is simulated in the present paper using two different numerical simulation techniques, namely 2-Zone model using Ozone and Computational Fluid Dynamics using Fire Dynamics Simulator (FDS). Through the FDS model, several features of the test can be thoroughly modeled to increase the accuracy of the results, however increasing the calculation time. While the 2-Zone model delivers quick and accurate results (time versus temperature development) especially if the model can be calibrated by the use of tests or based on visual observations reported by the fire brigade during the fire.

Keywords: FDS · 2-Zone model · Single story · Real fire · Post-fire assessment

1 Introduction

Post-fire load bearing assessment of industrial structures is of high economic importance for most companies, which must deal with such accidental event. In many cases, the structure exposed to the fire still stands and can be reused. And, therefore, the financial as well as time consequences for the company can be reduced. To obtain a better insight in the temperature development of load-bearing elements reached by the fire, there is a need for fast and reliable predicting models further evidenced by observations made during and after the fire by, for instance, the fire brigade. In the perspective of the structural response, pyrolysis with a gasification stage followed by

P. A. G. Piloto et al. (Eds.): CILASCI 2019, LNCE 1, pp. 48–60, 2020.
https://doi.org/10.1007/978-3-030-36240-9_4

combustion is of no interest. Nevertheless, the assessment should at least provide insight into the real temperature development that took place, based on a limited set of data.

This paper deals with post-fire situations, hence the fire-load can be well described which is most of the time not the case, especially in a design. The knowledge of the thermal load on a structure and its initial boundary conditions [1] are the main parameters to start with a resistance verification in case of fire. To obtain an idea of the temperatures reached in the structural elements during and after the fire, measurements (based on tests), photos and videos can be of great help but, most of the time, only some external visible parameters can be used, like, for instance, the melting or degradation temperature of the materials included in the building. The accuracy of such an evaluation is high enough when the considered material is fully accessible and not influenced by the surrounding environment. On the contrary, the temperature reached by non-visible parts of the structure (like the upper flange of a beam for example) can rarely be estimated. To fill this lack of knowledge, a simulation of the effect of a real fire on a structure can be done by the for example 2-Zone models or using Computational Fluid Dynamics (CFD). Validation can be made using the collated (visual) information to evaluate the accuracy of the model.

2 Reference Full-Scale Fire Tests

Only very few well instrumented and controlled fire tests on buildings are available to study the accuracy of simulations. On several occasions, the carried test did not lead to the expected conclusions because of environmental parameters. For example, during a previously executed real scale fire test in Belgium, in collaboration with KU Leuven, it was not possible to foresee the location of collapse due to a change in the wind direction during the test [2]. In France, a warehouse was set into fire in the scope of the National project Flumilog [3] and, there too, a strong influence of the wind was noticed.

Recently, new data on the experimental study of a full-scale steel portal frame submitted to a real fire were published in [4] and [5]. This study was mainly done in collaboration with Tongji university, which is the reason why we will name it the Tongji-experiment in the present paper. Photos of the tested building were taken each 5 min, hence there is a possibility to compare our simulations against visual observations during the fire. Gas temperature data at 4 locations inside the compartment were also registered. Thermocouples were placed at four levels on columns, beams and rafters. All the measurements were published in [4]. The fire load was obtained via wooden cribs and is therefore supposed to be well-known. Based on this, a model of the test was prepared using an Fire Dynamics Simulator (FDS) version 6.7 software [6] to simulate the effect of the fire on the building. In this article, the focus will be put on how the combustion parameters can be obtained in a rather straightforward way and on the results of an advanced pyrolysis study.

3 The Tongji Experiment

3.1 Geometrical Information

An extensive description of the experiment and the recorded measurements can be found in [4] and [5], some data needed to understand the test will be described here below and illustrated in Fig. 1(a). Two steel frameworks were erected with a span of 12 m, a roof eave at 5.4 m and a roof ridge at 5.8 m. The building skin was made of sandwich panels i.e. rock wool insulation core in between two thin-walled steel sheets. An inner partition wall was added with a fire rate of 3 h (ISO834 fire), with an opening along the edges of 0.3 m on top and 0.15 m on the sides as well as a door of 1.0 by 1.8 m. By doing so, a fire compartment of 4.0 by 6.0 m is created. There is one window opening of 1 m wide and 0.8 m height. The authors do not provide information on the type of window that was used hence single glazing will be considered as well as the following assumption (based on the Luxemburg rules for Fires Safety Engineering purposes [7]): in combination with a smoke temperature of 100 °C or 250 °C, the window opening reaches 50% and 90%, respectively. It is worth pointing that, at this stage, it was already demonstrated in [8] that the stress distribution in the glass does not fluctuate a lot between 250 °C and 400 °C, and therefore neither does the failure risk. In the compartment next to this one, one external door of 2.0 by 3.0 m was included.

Another important statement at this stage concerns the opening factor O of the compartment which will be rather small. At the beginning of the fire, when the glazing is still intact, O is given by:

$$O = \frac{A_v\sqrt{h_{eq}}}{A_t} = \frac{1 \cdot 1.8\sqrt{1.8}}{2 \cdot 4 \cdot 6 + 2 \cdot 5.6 \cdot (4 + 6)} = 0.015 \tag{1}$$

Where A_v = area of vertical openings, A_t = total area of enclosures and h_{eq} the weighted average of the window heights = $(\Sigma A_{vi}h_i)/\Sigma A_{vi}$.

The factor O is lower than the minimum boundary of 0.02 recommended in [9]. However, due to the glass breaking, the window opening reaches 90% of its theoretical surface, hence the opening factor increases up to 0.02 after this phenomenon. Equation 1 becomes Eq. 2 and it can be presumed that the fire will be ventilation controlled:

$$O = \frac{A_v\sqrt{h_{eq}}}{A_t} = \frac{(1 \cdot 1,8 + 1 \cdot 0.8)\sqrt{1.49}}{2 \cdot 4 \cdot 6 + 2 \cdot 5.6 \cdot (4 + 6)} = 0.020 \tag{2}$$

Thermocouples were placed on trees (= vertical steel bar with thermocouple devices attached at different levels) between the wooden piles as well as on the steel elements (locations 1# and 4#), where 4 positions were measured: at 1/4th 2/4th, 3/4th and on top of the element; in the neighbouring compartment (locations 2# and 3#), where 2 positions were measured: at half and on top of the element.

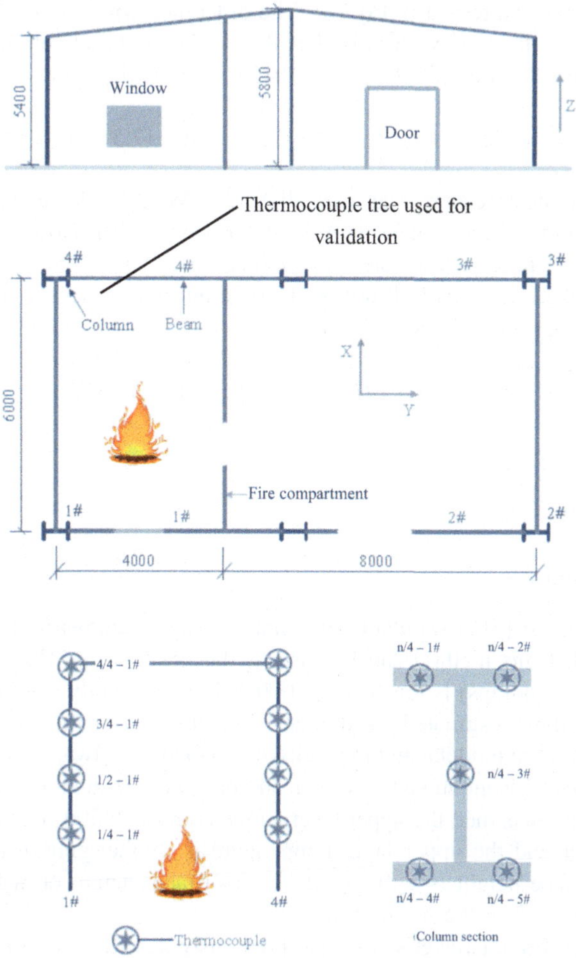

Fig. 1. Overview of the structure and fire compartment (a) and measurement equipment (b) [5]

3.2 Fire Load

Following the description, a fire load of 81900 MJ was present. It was made of four piles of wooden cribs. Nevertheless, our calculation revealed an overprediction of that load, which was, in the meantime, confirmed by the authors themselves. Indeed, four piles of wooden cribs were present, each pile made of 20 layers with 15 sticks on each layer. Depending on the layer, the wooden lengths were ranging between 1.5 and 2 m, with a cross-section of 50 by 50 mm^2. Considering a combustion heat for wood of 17.5 MJ/kg and a mass density of 440 kg/m^3, we obtain half of the fire load provided in [4]. This can be worked out as the product of [the number of sticks] by [half of the number of layers] by [the length of parallel and perpendicular layers] by [the gross section] by [density] by [combustion heat] as expressed in Eq. (3). The difference with the reference article is that the authors used the gross volume of the piles instead of its net volume. With a floor area of only 24 m^2, the fire load density is 40425/24 = 1684 MJ/m^2.

$$\text{Fire load} = 4 \cdot 15 \cdot 10 \cdot (1.5 + 2) \cdot 0.05^2 \cdot 440 \cdot 17.5 = 40425 \text{ in } [\text{MJ}] \quad (3)$$

The measured heat release rate was about 5 MW per pile, a similar value is obtained by the application of the annex E of Eurocode 1 [9]. However, in [4], it is stated that the total heat release rate was (4 times 5) 20 MW. Nevertheless, with a ventilation-controlled fire, this will not be the case and the prescriptions of the Eurocode 1 limit the heat release rate to Eq. (4):

$$\text{HRR} < 0.10 \cdot m \cdot H_u \cdot A_v \cdot \sqrt{h_{eq}} = 4.45 \text{ in[MW]} \quad (4)$$

4 Fire Models

4.1 2-Zone Model

In the reference article [4], a simulation was made using a parametric fire method. It is worth pointing that this method can be used as the fire load $q_{t,d}$ (based on the total surface of the surroundings) is lower than 1000 MJ/m^2. Presently it is 1684 MJ/m^2 of floor surface (initially estimated as double). For this reason, a simulation of the experiment as a 2-Zone model using the well-known Ozone software version 3.0.1 [10] was carried out. The 2-Zone model automatically becomes a 1-zone model if one of the following four criteria is met: the upper layer temperature is ≥ 500 °C; the combustible in the upper layer and the upper layer temperature is $\geq$ the ignition temperature of 300 °C; the interface height is ≤ 0.2 of the compartment height or at least the combustion area is $\geq 1/4^{th}$ of the floor surface.

A user-defined fire characterised by the previously described geometry of the fire compartment was introduced, with a t_α of 300 s, a heat release rate of 500 kW/m^2, a fire load of 1684 MJ/m^2 and a danger of fire activation equal to 1. The heat release rate (HRR) and the temperature of the upper layer (Ozone) are respectively presented in Figs. 2 and 4.

A comparison between the visual observations during the test and those preliminary calculations was also made. It shows that 5 min after ignition, the fire was seriously developing but that the glazing of the window was still intact. In other words, the flashover occurs 300 s after ignition, despite the intact glazing. The Ozone calculation leads to a flashover after only 60 s seen that, at that time, it automatically switched from 2 to 1 zone.

As for the post-fire simulations, it can be noticed from Figs. 2 and 4 (Ozone shifted), where the Ozone upper layer temperatures and the measured steel temperatures in the column 1# are provided, that the tool is appropriately predicting the smoke temperature. Note that, the Ozone temperature-time relation is shifted by 500 s which is the ignition time and that the smoke temperature is compared to the steel temperature causing a

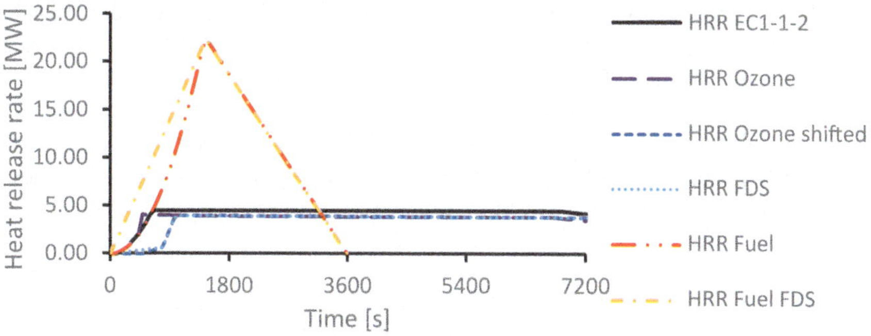

Fig. 2. HRR obtained by Ozone, application of EN1991-1-2 or a fuel-controlled fire

slight decrease. The visual observations become rather important to define this so-called shift. Indeed, since glazing is still intact after 800 s (ignition time + 300 s), the smoke temperature at that moment should still be around 250 °C i.e. the failure criteria for glass. Using Ozone, this happens after 286 s.

On the one hand, the smoke temperature calculated by Ozone is slightly in advance compared to the steel temperature (which may be expected). But, on the other hand, this is not the case for the thermocouple situated at 3/4th of the column height. After 900 s or 15 min of fire (= after ignition), a peak temperature over 1100 °C is measured as can be seen in Fig. 4.

Moreover, the structure starts to fail, the wall cladding tears apart leading to an extra amount of oxygen at that location with higher HRR and temperatures. It is also clear that, after 19 min of fire, the shape and volume of the compartment have

Fig. 3. Visual observations during the test [5]

drastically changed, see Fig. 3. Due to the extra amount of oxygen and the changes in shape, the interesting time range is limited to about 15 min and definitely lower than 19 min after fire ignition [5]. It is the area shaded in light grey in Fig. 4.

After a thorough study of the articles [4] and [5], which both describe the same experiment, it came out that the reported measured steel temperatures beneath 800 s were different. After feedback of the authors, it became clear that the results as they are presented in the journal article are more reliable.

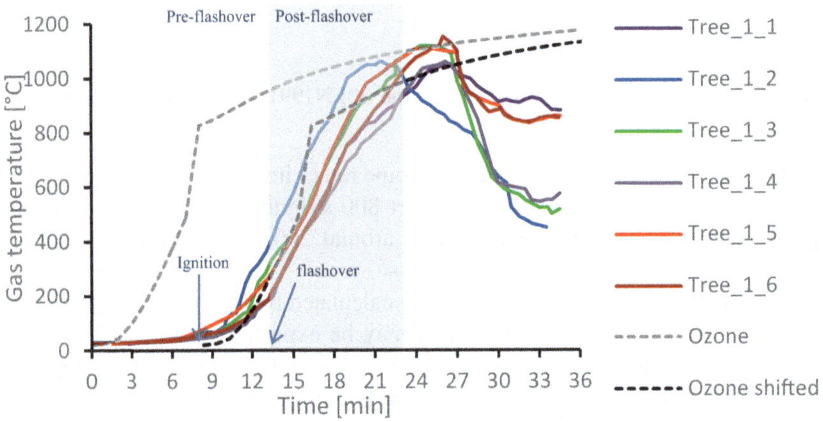

Fig. 4. Measured temperature development on the steel Sect. 1# [5] together with the Ozone evaluation of the upper layer temperature

4.2 FDS Model

FDS or Fire Dynamics Simulator [6] is a widely used powerful tool for the description of the fluid behavior in case of fire. Through the use of computational fluid dynamics, it is possible to go much further in detail in the simulation but on the other hand this numerical tool requests also much more information to describe the fire and the boundaries.

General Introduction to the Model. The smallest dimension of the wooden cribs is 50 mm. For that reason, the minimum mesh size was put equal to 50 mm. A mesh sensitivity check was executed with a finer mesh of 25 mm cubes till 15 min, the comparison will be shown later on in this section. The FDS version 6.7 [6] was used. The numerical domain was limited to the combustion area except for the wall with an opening, where an extension of 1.5 m outside was considered. The geometry of the model is shown in Fig. 5.

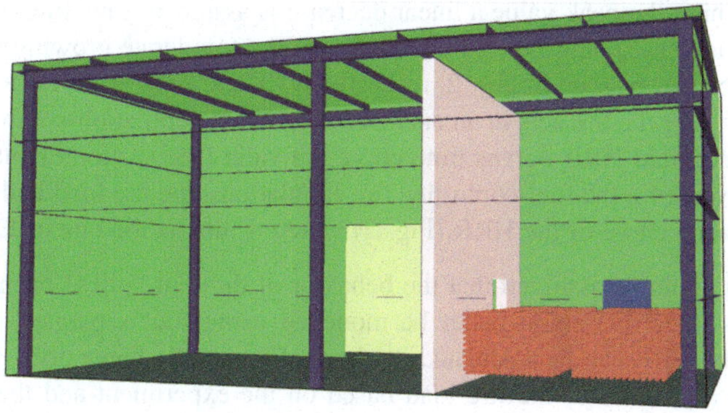

Fig. 5. FDS Model of the structure, with view on fire load, window, door and gate

The floor is made of concrete, sandwich elements are used for the walls, the roof and the open space for the external part. Standard material characteristics were collected out of the literature as presented in Table 1 as they are not provided in [5].

Table 1. Material characteristics

Description	Material	Thickness [m]	Conductivity [W/mK]	Specific heat [J/kgK]	Density [kg/m^3]	Backing
Floor	Concrete	0.15	1.6	1000	2300	Exposed
Fire wall	Cellular concrete	0.20	0.14	840	550	Exposed
Profile	Steel	0.01	45	600	7850	Exposed
Sheet	Steel	0.0005	45	600	7850	Exposed
Insulation	Wool	0.075	0.037	1030	60	Exposed

Applied Fire Load. The global physical reaction of the wooden cribs was introduced as $C_{3.4}H_{6.2}O_{2.5}$ with a soot yield of 0.015 and a heat of combustion equal to 17.5 MJ/kg. No attention was paid to travelling fire due to the relatively small size of the compartment.

Uniformly Distributed Fire Load. As first simulation was used as upper bound: the combustion product develops the highest possible HRR. The combustible amount of energy was previously calculated as 40425 MJ. In the assumption of a linearly increasing heat over a period of 1440 s (time of maximum reached temperature in the compartment) and decreasing over 2160 s, the energy over time is 22458 kW. This is more or less the description of a fuel-controlled fire with a quadratic growing stage as mentioned in [9] with the formulation of Eq (5).

$$Q = \left(\frac{t}{t_\alpha}\right)^2 \text{ in [MW]} \tag{5}$$

After reaching the peak value a linear decrease is assumed. The difference between the quadratic increase using Eq. (5) with a $t_\alpha = 300$ s (moderate growing rate) and the linear approach can be seen in Fig. 2, respectively "HRR Fuel" and "HRR Fuel FDS".

In sum, as long the fire acts in an enclosed space, it is ventilation-controlled and the previously calculated HRR versus time curve are valid (see section about the 2-Zone model). So, except for the fuel-controlled fire simulations, the fire load will be based on the shifted Ozone curve of the HRR (Fig. 2) which is almost the one of EN 1991-1-2.

Local Fire. It could be assumed that the behavior of the four "piles" of interlaced and layered sticks (or cribs) could better be modelled using four separate local Hasemi-fires. In [9], the HRR can be calculated on the roof where the flame does not touch the ceiling. With a position of the fire load based on the experiment and the FDS simulation (Fig. 5), a quasi-uniform HRR can be obtained at the roof level. For the size of the fire load, an equivalent diameter of 1.96 m was calculated. The Fire height is taken equal to 1 m from the top of the wood pile.

4.3 Parameters Investigated in the FDS Model

Geometry of the Fire Load. Based on [11], where a simplified predefined heat release rate pro unit area (HRRPUA) is provided, several simulations were presently done, in which the geometry of the fire together with the condition which control the fire (fuel or ventilation) leads to different HRRPUA are as follows:

- For a crib under the condition of a fuel-controlled fire, we have: $22458/[4 \cdot 15 \cdot 10 \cdot (1.5 + 2.0) \cdot 4 \cdot 0.05] = 54\ \text{kW/m}^2$, since cribs are presumed to burn on four sides.
- For a crib with a ventilation-controlled fire: $4450/[4 \cdot 15 \cdot 10 \cdot (1.5 + 2.0) \cdot 4 \cdot 0.05] = 10.60\ \text{kW/m}^2$, under the same conditions as above.
- Pro pile for a ventilation-controlled fire: $4450/[4(1{,}5 \cdot 2{,}0 + 2 \cdot (1.5 + 2.0) \cdot 1)] = 111{,}25\ \text{kW/m}^2$, it is assumed that the burning face corresponds to the top and the four sides.

Window Breaking. In Sect. 3.1, it was shown that O is rather low and could impact the fire development. For that reason, several fire glass-breaking scenarios were investigated:

- With no glass at all (upper bound), refer to the last section about the thermocouple devices.
- With the glass consisting of an upper (from 1.4 till 1.8 m) and lower part (from 1.0 to 1.4 m) and breaking at a temperature of 250 °C (the temperature being controlled at 0.05 m out of the glass (to the fire).
- Same as above. At this moment no longer a simple temperature device is used but a thermocouple device. The difference between these two will be discussed in the last section about the thermocouple devices.

Mesh Size. For the models with cribs of 5 by 5 cm, a mesh size of 5 cm in all directions is used which can be doubled or tripled (going up to 0.15 m) when a pile is considered. The latter value was kept as the maximum value due to the size of the roof girders (the height of which is 0.15 m). In the simulation, a perfect cubic mesh geometry was used.

Thermocouple Devices. Since no specification about the used devices could be found in [5], traditional temperature devices measuring the temperature of the gases were firstly included in the first two models. In [4] however, it is mentioned that k-type thermocouples were used thermocouples were then used in the subsequent simulations. The default setting parameters for a FDS thermocouple were used (i.e. diameter 1 mm, emissivity 0.85, density 8909 kg/m^3 and specific heat 0.44 kJ/kg/K for nickel).

5 Results

Eight simulations were conducted in total, as can be found in Table 2. One can find, in order of appearance, the type of HRR: simplified model based on measurements of time and temperature (fuel-controlled) or the so-called shifted 2-Zone Ozone model (ventilation-controlled); the fire load: cribs or pile; the scenario of window openings: (a) completely open from the start, (b) one glass panel or (c) failure in two steps; the type of temperature measurement devices: T = Temperature and TC = thermocouple device; and, last, the mesh size (Mesh) in [m]. Three time-related information are provided too; the simulation time (stopped after about 35 min), the average of the achieved simulation time in seconds per day and the total simulation time needed in days.

Table 2. Overview of FDS models

#	HRR	Fire	Window	Dev.	Mesh [m]	Time [s]	CPU [s/day]	CPU [days]
1	Meas.	Cribs	(a)	T	0,05	2153	9	227
2	Ozone	Cribs	(c)	T	0,05	2131	23	93
3	Ozone	Pile	(c)	T	0,05	2131	27	80
4	Ozone	Pile	(b)	TC	0,05	2117	22	98
5	Ozone	Pile	(b)	TC	0,1	2100	263	8
6	**Ozone**	**Pile**	**(a)**	**TC**	**0,1**	**2100**	**263**	**8**
7	Ozone	Pile	(b)	TC	0,15	2100	2100	1
8	Meas.	Pile	(b)	TC	0,15	2135	1068	2

The first model cannot capture in a proper the delay of the ignition or the temperature development because of the limited ventilation conditions which are presently not properly modelled. In the second model, thanks to the shifted HRR, a better approximation of the start of the fire is obtained. However, the opening scenario (c) leads to heavily fluctuating temperatures. Simulating the fire load using solid volume instead of bricks delivers better results. The temperatures are well predicted till about 500 °C. Changing to one glass panel does not influence the results. The same can be concluded when increasing the mesh size to 0,15 m. It should be noted that if the HRR is calculated with glass breaking, it is redundant to include the window in the simulation.

Hence, the next simulation, which is the one depicted in Fig. 6, was done with an open window (scenario (a)) from the start. In this figure, the measured temperatures of tree #1 are provided together with the original and shifted 2-Zone Ozone simulation in grey and black dashed lines, respectively. The FDS calculated gas or thermocouple temperature are provided as dotted lines, for the simulation #6 (in bold).

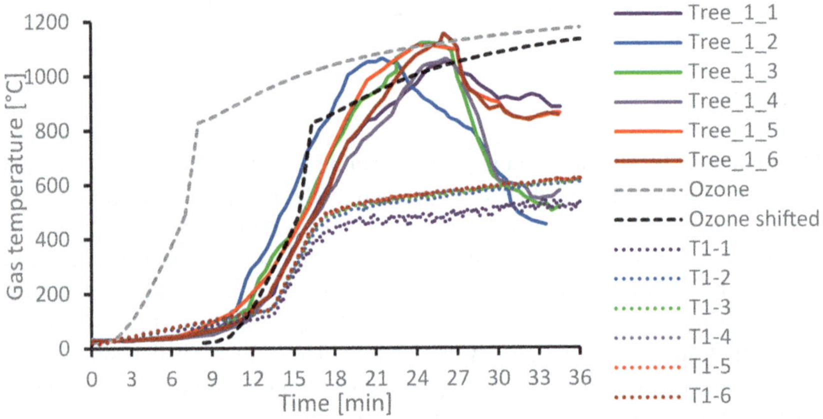

Fig. 6. Temperature measurements, Ozone and FDS for ventilation-controlled, pile fire without glass panel and 0.10 m mesh size

Out of Fig. 1, one can notice that FDS, 2-Zone Ozone and the experiment itself deliver comparable results until about 450 °C. The simulation #6, though delivering the best agreement with the measured results, is however not able to capture the post-flashover behavior in an accurate way. The obtained temperatures remain at about 400 °C too low compared to the test or to the 2-Zone Ozone results. Solving energy and mass equations for such a large compartment (compared to the cell size) is probably causing the large discrepancy, however this should be properly and more thoroughly investigated.

6 Conclusions

This paper describes a quick methodology to obtain the temperatures reached in a structure during a fire and demonstrates its accuracy with a case-study. The case study is a steel frame single-story building submitted to a real fire. It was simulated in the present paper using two different numerical simulation techniques, namely 2-Zone model using Ozone and Computational Fluid Dynamics using Fire Dynamics Simulator (FDS).

Today, even though FDS is considered inappropriate to simulate flashover in large compartments, it still stays a powerful instrument to study more local effects. As such it can deliver valuable and helpful results for post-fire assessment. Through the FDS model, several features of the test can be thoroughly modeled to increase the accuracy of the results, however increasing the calculation time. Despite the sometimes immense required computing time, one herein demonstrated that the FDS model could predict the temperature development in a very accurate way, however, in this case, till a relatively limited temperature of about 450 °C. One of the simplest FDS model delivers the best fit with the measured temperatures, i.e. using a solid burning volume and the least fine mesh. As mentioned in the paper, since the calculated heat release rate took the glass breaking into account, all openings were set open from the beginning of the simulation, without breaking scenarios.

The 2-Zone model delivers quick and accurate results (time versus temperature development) especially if the model can be calibrated by the use of tests or based on visual observations reported by the fire brigade during the fire. The results are filling the gaps between what can be easily observed after the fire (or during, by the fire brigade) and what remains unknown. Looking towards time efficiency and accuracy, the 2-Zone model performed well, even till collapse of the building.

For post-fire assessment purposes, it can therefore be concluded that, only under very particular circumstances, an advanced modelling of the fire and the compartments will have to be carried out. A simpler model, such as the 2-Zone model presented in this paper can advantageously be used. However, a number of information are required to calibrate the model such as the temperature attained by some elements of the structure for example as well as the heat release rate per unit area. In the Tongji experiment, a big amount of data was available but, in most cases, only a few information will be available, and one has to base the iteration process only on a few specific known times (like failure or flashover) or temperatures. Herein, the time of glass breaking was, in this specific case-study, an important parameter used to validate the simulation.

References

1. Merci, B., Beji, T.: Fluid Mechanics Aspects of Fire and Smoke Dynamics in Enclosures. CRC Press, Leiden (2016)
2. Pyl, L., Schueremans, L., Dierckx, W., Georgieva, I.: Fire safety analysis of a 3D frame structure based on a full-scale fire test. Thin-Walled Struct. **61**, 204–212 (2012)
3. Thauvoye, C., Russo, P., Blanchet, J.-M., Duplantier, S., Kruppa, J., Muller, A., Patej, S., Taveau, J., Zhao, B.: Method for calculating heat fluxes from a warehouse fire. In: 8th International conference on Performance-Based Codes and Fire Safety Design Methods, Lund (2010)
4. Lou, G., Wang, C., Jiang, J., Jiang, Y., Wang, L., Li, G.-Q.: Experimental and numerical study on thermal-structural behavior of steel portal frames in real fires. Fire Saf. J. **98**, 48–62 (2018)
5. Zhong, B., Jiang, Y.-Q., Lou, G.-B.: Experimental study on the thermal and structural responses of a full-scale steel structure under natural fire. In: The 10th International Conference on Structures in Fire, Belfast (2018)

6. McGrattan, K., Hostikka, S., McDermott, R., Floyd, J., Vanella, M.: Fire Dynamics Simulator; User's Guide. NIST & VTT, Maryland (2018)
7. EN 1991-1-2:2002/AN-LU:2011, Luxemburg national annex to Eurocode 1: Actions on Structures - Part 1-2: General actions - Actions on structures exposed to fire, Luxembourg: ILNAS (2011)
8. Wang, Y., Wang, Q., Shao, G., Chen, H., Sun, J., He, L., Liew, K.: Experimental study on critical breaking stress of float glass under elevated temperature. Mat. Des. **60**, 41–49 (2014)
9. EN 1991-1-2, Eurocode 1 - Actions on Structures - part 1-2: General actions - Actions on structures exposed to fire, Brussels: CEN (2002)
10. Cadorin, J.-F., Pintea, D., Dotreppe, J.-C., Franssen, J.-M.: A tool to design steel elements submitted to compartment fires - Ozone v2 Part2: methodology and application. Fire Saf. J. **5**, 429–451 (2003)
11. Degler, J., Eliasson, A., Anderson, J., Lange, D., Rush, D.: A-priori modelling of the Tisova fire test as input to experimental work. In: ASFE, Dubrovnik (2015)

Buckling Resistance of Partially Encased Columns Embedded on Walls Under Fire from One Side

Paulo A. G. Piloto[1], Nathália Gonçales[2](✉), Ronaldo Rigobello[2], Mário Vaz[3], Rui M. Guedes[3], and João S. Baptista[3]

[1] Instituto Politécnico de Bragança, Bragança 5300-253, Portugal
[2] Universidade Tecnológica Federal do Paraná, Paraná 87301-899, Brazil
nathaliagoncales@alunos.utfpr.edu.br
[3] Faculdade de Engenharia, Universidade do Porto, Porto 4200-465, Portugal

Abstract. Partially encased columns (PEC) have better fire resistance when compared to bare steel columns, due to the existence of concrete between the flanges. The aim of this study is to develop a new proposal for the calculation of the buckling strength of partially encased columns, embedded on walls, under fire conditions. This proposal is based on the current calculation method proposed in Annex G of EN 1994-1-2. This study uses the finite element method to calculate the average temperature of seven components. The average temperature is then used to find the buckling resistance of composite columns when submitted to fire from one side. This solution method is carried out using 30 different cross sections. All cross sections are exposed to the standard fire curve ISO834 from one side, assuming the specific fire rating of 30, 60, 90 and 120 min.

Keywords: Partially encased columns · Fire resistance · Columns embedded on walls · Numerical simulation

1 Introduction

The steel columns have a great loadbearing capacity, however when exposed to fire conditions, they resist only to a relative short period of time. The partially encased composite columns consist in an excellent alternative solution to bare steel column. Partially Encased Columns (PEC) have higher strength and stiffness when compared to steel bare profiles under fire conditions. The fire resistance of the PEC depends on the temperature evolution in each material and component, being this temperature field affected by the protection effect of the wall.

The annex G of Eurocode EN 1994-1-2 [1], presents the balanced summation method, allowing the calculation of the buckling resistance of PEC, when submitted to standard fire conditions [2] from the 4 sides, but this annex does not take into consideration the embedded effect of this PEC in the wall. This method differs from the general calculation method and divides the cross section into four large components.

The non-symmetric temperature field over different types of materials and elements have been studied for long time. In 1988, Cooke [3] presented an experimental and

P. A. G. Piloto et al. (Eds.): CILASCI 2019, LNCE 1, pp. 61–77, 2020.
https://doi.org/10.1007/978-3-030-36240-9_5

theoretical study about the thermal bowing of building elements. The temperature field and gradient over the cross section depends on the material under analysis. Test data show that large thermal bowing deflections occur in unrestrained brick walls. The author also presented simple calculation models to deal with unrestrained thermal bowing effect on steel elements. The behaviour of composite elements is more complex due to the existence of non-linear thermal gradients in concrete. In 2007, Garlock and Quiel [4] presented the effect of thermal gradients in steel columns with wide flanges, which are responsible for modifying the position of the neutral axis to the coldest side, introducing extra bending moments. This shift effect also depends on the amount of axial load. In 2010, Correia et al. [5] presented a new proposal for the calculation method to evaluate the temperature of H steel columns embedded in single-leaf brick walls. This proposal was based on numerical simulations and fire resistance tests, taking into consideration two different relative positions between the steel profile and the brick wall. The proposal considered a temperature gradient, based on 3 isothermal zones or components (exposed, embedded and unexposed). In 2011, Dwaikat et al. [6] tested several wide-flanged steel beam-columns with induced temperature gradients. These temperature gradients were induced by the effect of the furnace into a partially protected column, considering the inexistence of fire protection materials in certain regions, allowing to define two main directions for the highest gradient (parallel and perpendicular to the web). This procedure simulated, according to authors, a realistic three-sided heating scenario. The column specimens developed bending moments in regards to these gradients. Major gradients were defined in the direction parallel to the web for both fire events. All columns failed by full section yielding. The plastic resistance due to the combinations of axial load and moment was affected by the thermal gradients. The experiments and computer models showed good agreement with the predicted demands and capacity. In 2014, Agarwal et al. [7] made an experimental and numerical investigation to analyse the behaviour of columns with thermal gradients in the cross section. The columns were submitted to uniform heating, that reached their failure temperatures faster than the columns subjected to non-uniform heating. The parametric analysis allowed the development of design equations for wide flanged steel columns subjected to non-uniform heating. In 2014, Quiel et al. [8] investigated three different models to predict the axial plastic load bearing capacity of steel columns with thermal gradients, using the code-based equations (American and European), using the fibre-beam element model and using the shell finite element model. Authors concluded that code-based equations are not satisfactory, since these equations do not consider temperature gradients. Both finite element models agree well with experimental results. The tests and models were very important to develop new simple formulae, which included de effect of moment reversal due to a shift in the section centre of stiffness, produced by the existence of temperature gradients. In 2014, Correia et al. [9] presented a numerical study that considered non-uniform temperature distribution in the cross section of restrained steel H columns embedded on walls, validated by experiments, proposing new interaction formulae for axial force and bending moment. The contact of the steel columns with the walls is responsible for a big reduction in the temperature of the cross-section, leading to higher fire resistance when compared to the ones observed in engulfed steel columns. According to the authors, these columns behave much more like beam-columns failing by bending, following the effect of the "thermal bowing", instead of failing by buckling. The application of the Eurocode 3 simple formula, considering a uniform temperature, within the cross-section, equal to the temperature of the exposed flange, would lead to a very conservative design. In 2016, Ojeda et al. [10] studied the effect of thermal gradients over the cross-section of steel columns by means of finite element simulations. Authors investigated the eccentricity in the column, created by the temperature gradient and the reduction of the flexural buckling resistance of the columns. Authors decided to simplify the calculation of the required parameters in order to handle a simple calculation method. A new design model was proposed, consisting of a set of simple equations which considers the eccentricity. The simplifications involve the calculation of the required material properties and geometric parameters at the average, maximum or minimum temperatures in the section. Rocha et al., in 2018 [11] studied the fire behaviour of steel columns and PEC embedded on masonry walls with restrained thermal elongation. Authors found that the magnitude of the thermal gradients and their directions have a strong effect on the mechanical behaviour and stability of the PEC under fire. The thickness of the wall influenced the bending stiffness of the tested columns, affecting the restraining forces and displacements. The thermal gradient in the cross section is responsible for the thermal bowing effect, introducing additional bending moments and axial forces.

This study presents an new approximation method to define the axial buckling resistance of Partially Encased Columns, embedded on walls, under fire, taking into consideration the temperature effect over the plastic resistance to axial compression $N_{fi,pl,Rd}$ and over the effective flexural stiffness $(EI)_{fi,eff,z}$ around the minor axis.

2 Partially Encased Columns Embedded on Walls

This work investigates partially encased columns embedded on clay walls under standard fire from only one side (see Fig. 1) and proposes an alternative method to the balanced summation model, including the contribution of more components, due to the existence of asymmetric temperature field, with respect to the plane defined by the web of the PEC.

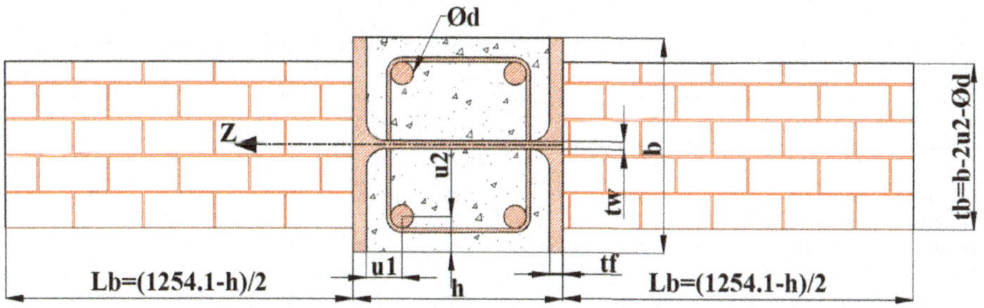

Fig. 1. Partially encased column embedded on wall.

The thermal analysis of PEC deals with 30 different PEC, made by HEA, HD and UC cross section. The length and thickness of the wall changed, according to the values presented on Table 1.

Table 1. Section proprieties for the partially encased sections.

Profiles	h [mm]	b [mm]	tw [mm]	tf [mm]	Am/V [m^{-1}]	Nº of bars	ϕ [mm]	As/As + Ac [%]	u [mm]	Rebar area [mm^2]	Wall length (Lb) [mm]	Wall thickness (tb) [mm]
HEA 240	230	240	7.5	12	6.30	4	20	2.62	50.0	314.2	512.05	122.28
HEA 280	270	280	8	13	5.19	4	25	2.96	50.0	490.9	492.05	157.84
HEA 320	310	300	9	15.5	4.65	4	25	2.42	50.0	490.9	472.05	177.84
HEA 360	350	300	10	17.5	4.56	4	32	3.52	50.0	804.2	452.05	171.64
HEA 450	440	300	11.5	21	4.31	4	32	2.80	50.0	804.2	407.05	171.64
HEA 500	490	300	12	23	4.21	4	32	2.52	50.0	804.2	382.05	171.64
HEA 600	590	300	13	25	4.06	4	32	2.08	54.8	804.2	332.05	171.64
HEA 700	690	300	14.5	27	3.99	4	40	2.77	54.8	1256.6	282.05	164.55
HEA 800	790	300	15	28	3.90	4	40	2.40	54.8	1256.6	232.05	164.55
HEA 1000	990	300	16.5	26	3.79	4	40	1.89	54.8	1256.6	132.05	164.55
UC 254 x 254 x 73	254.1	254.6	8.6	14.2	5.82	4	25	3.54	50.0	490.9	500.00	132.44
UC 254 x 254 x 167	289.1	265.2	19.2	31.7	5.36	4	25	3.54	50.0	490.9	482.50	143.04
UC 305 x 305 x 97	307.9	305.3	9.9	15.4	4.57	4	25	2.40	50.0	490.9	473.10	183.14
UC 305 x 305 x 158	327.1	311.2	15.8	25	4.41	4	25	2.40	50.0	490.9	463.50	189.04
UC 305 x 305 x 313	374	325	30	48.3	4.08	4	25	2.40	50.0	490.9	440.05	202.84
UC 356 x 368 x 202	374.6	374.7	16.5	27	3.58	4	32	2.80	50.0	804.2	439.75	246.34
UC 356 x 406 x 340	406.4	403	26.6	42.9	3.27	4	32	2.67	50.0	804.2	423.85	274.64
UC 356 x 406 x 592	465	421	45	72.3	3.03	4	32	2.67	50.0	804.2	394.55	292.64
UC 356 x 406 x 744	498	432	55.6	88.9	2.91	4	32	2.67	50.0	804.2	378.05	303.64
UC 356 x 406 x 1086	569	454	78	125	2.70	4	32	2.68	50.0	804.2	342.55	325.64
HD 260 x 93,0	260	260	10	17.5	5.65	4	25	3.49	50.0	490.9	497.05	137.84
HD 260 x 172	290	268	18	32.5	5.30	4	25	3.49	50.0	490.9	482.05	145.84
HD 320 x 74,2	301	300	8	11	4.69	4	25	2.41	50.0	490.9	476.55	177.84
HD 320 x 198	343	306	18	32	4.43	4	25	2.44	50.0	490.9	455.55	183.84
HD 320 x 300	375	313	27	48	4.24	4	25	2.46	50.0	490.9	439.55	190.84
HD 400 x 237	380	395	18.9	30.2	3.39	4	32	2.68	50.0	804.2	437.05	266.64
HD 400 x 347	407	404	27.2	43.7	3.26	4	32	2.67	50.0	804.2	423.55	275.64
HD 400 x 509	446	416	39.1	62.7	3.10	4	32	2.66	50.0	804.2	404.05	287.64
HD 400 x 900	531	442	65.9	106	2.81	4	32	2.68	50.0	804.2	361.55	313.64
HD 400 x 1299	600	476	100	140	2.55	4	32	2.67	50.0	804.2	327.05	347.64

The grades of the materials are: S355 for the steel profiles, C20/25 for the encased concrete and B500 for rebars. A clay wall was considered on both sides of the partially encased section. The cross-section dimensions were selected to present a wide range of variation for the value of the section factor. The exposed perimeter and the total area of the cross section were used to calculate the section factor (see Eq. 1).

$$A_m/V = [(b - tb) + h]/(b\,h) \tag{1}$$

The balanced summation model, presented in annex G [1], was originally developed to determine the loadbearing capacity of PEC under fire, dividing the cross section in four components, based on the assumption of doubly symmetric thermal behaviour and fire from 4 sides. In the following formulae, the four components of PEC are identified with, "f" for flanges, "w" for web, "c" for concrete and "s" for the reinforcing bars. The effect of each component on the plastic resistance to axial compression can be calculated using Eq. 2 (normal summation) and the effect on the effective flexural stiffness determined by Eq. 3 (balanced summation with reduction coefficients).

$$N_{fi,pl,Rd} = N_{fi,pl,Rd,f} + N_{fi,pl,Rd,w} + N_{fi,pl,Rd,c} + N_{fi,pl,Rd,s} \tag{2}$$

$$(EI)_{fi,eff,z} = \phi_{f,\theta}(EI)_{fi,f,z} + \phi_{w,\theta}(EI)_{fi,w,z} + \phi_{c,\theta}(EI)_{fi,c,z} + \phi_{s,\theta}(EI)_{fi,s,z} \tag{3}$$

In the current version model, the average flange temperature is obtained through equations using empirical factors, depending on the section factor. The yield stress and elastic modulus are affected by temperature, using reduction factors. The web geometry reduction is based on empirical factors. The yield stress is reduced using an indirect parameter, leaving the elastic modulus not affected by temperature. The concrete temperature is calculated using a table based on the section factor, for each class of fire resistance. Part of the concrete geometry is neglected using the same distance in the both principal directions and the mechanical proprieties are affected by temperature. The reinforcing bars have their mechanical properties affected by temperature, using the reduction factors for each size of concrete cover layer "u", obtained by the geometric average of "u_1" and "u_2". The material safety factors $\gamma_{M,fi,a}$, $\gamma_{M,fi,c}$ and $\gamma_{M,fi,s}$ for the structural steel, concrete and reinforcing steel, respectively, are assumed equal to 1. The reduction coefficients of balanced summation are presented in Table 2.

Table 2. Reduction coefficients for balanced summation.

Reduction coefficients for bending stiffness				
Fire rating	$\phi_{f,\theta}$	$\phi_{w,\theta}$	$\phi_{c,\theta}$	$\phi_{s,\theta}$
R30	1,0	1,0	0,8	1,0
R60	0,9	1,0	0,8	0,9
R90	0,8	1,0	0,8	0,8
R120	1,0	1,0	0,8	1,0

3 Buckling Resistance

The annex G of Eurocode 4 part 1–2 [1] presents the simple calculation method to determine the buckling resistance of PEC. This method considers the effect of the fire over four components (flanges, web, concrete and reinforcing bars), when the PEC is submitted to a fire scenario from 4 sides. The stability of PEC requires the calculation of the critical load, the plastic resistance to axial compression and the effective flexural stiffness.

In order to consider the effect of a different fire scenario (fire from one side), seven components are required.

The elastic buckling load $N_{fi,cr,z}$ requires the calculation of the effective flexural stiffness $(EI)_{fi,eff,z}$. The non-dimensional slenderness ratio λ_θ, depends on the plastic resistance to axial compression under fire $N_{fi,pl,Rd}$ and should be calculated according to Eqs. 4 and 5. The buckling length of the column under fire conditions is represented by L_θ.. In order to evaluate the buckling resistance, three buckling lengths were considered $L_\theta = 1L$, $L_\theta = 0{,}7L$ and $L_\theta = 0{,}5L$, assuming two different columns length (3 and 4 m). According to the solution method, the buckling curve "C" of EN 1993-1-1 should be used to find the buckling resistance [12], using the reduction coefficient χ_z.

$$\overline{\lambda_\theta} = \sqrt{N_{fi,pl,Rd}/N_{fi,cr,z}} \tag{4}$$

$$N_{fi,cr,z} = \pi^2 (EI)_{fi,eff,z}/L_\theta^2 \tag{5}$$

The Fig. 2 presents the flowchart of the simplified calculation method used for the calculation of the buckling resistance of PEC.

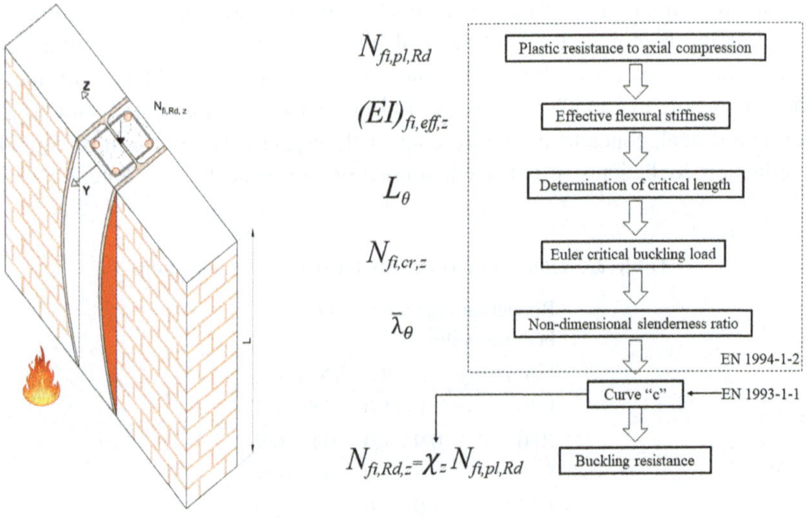

Fig. 2. Simplified calculation method.

4 Numerical Simulation

The purpose of this work is to perform numerical simulations of composite steel-concrete columns in contact with clay walls under fire exposure from one side. The advanced calculation method was developed using the finite element method. This method requires the solution of Eq. 6 when submitted to the boundary conditions Eq. 7. The solution method is incremental and non-linear, which requires an iterative process. The post processing of the temperature field over the PEC is used to determine the average temperature of each component (exposed/unexposed flange, web, exposed/unexposed rebars, residual areas of the exposed and unexposed concrete) for different fire ratings.

$$\nabla\left(\lambda_{(T)} \cdot \nabla T\right) = \rho_{(T)} \cdot C_{p(T)} \cdot \partial T/\partial t \rightarrow (\Omega) \tag{6}$$

$$\lambda_{(T)} \cdot \nabla T \cdot \vec{n} = \alpha_c\left(T_g - T\right) + \Phi \cdot \varepsilon_m \cdot \varepsilon_f \cdot \sigma \cdot \left(T_g^4 - T^4\right) \rightarrow (\partial\Omega) \tag{7}$$

In these equations: $\lambda_{(T)}$ represents the thermal conductivity, $\rho_{(T)}$ defines the specific mass, $Cp_{(T)}$ defines the specific heat, T represents the temperature of each material, T_g defines the gas temperature of the fire compartment, α_c defines the convective coefficient, ϕ defines the view factor, ε_f and ε_m define the emissivity of fire and material respectively and σ represents the Stephan-Boltzmann constant.

The finite element model, considers the incremental and iterative solution, using 2D finite elements PLANE55, which presents four nodes with a single degree of freedom (temperature at each node), using linear interpolating functions and full integration Gauss scheme. The mesh was well refined (0.5 mm maximum size) to cnable the calculation for the reduction of the concrete affected by the fire (exposed side) (see Fig. 3).

Fig. 3. Finite element mesh used for simulations.

Nonlinear thermal properties are used for steel, concrete and masonry, according to the Eurocodes EN1993-1-2 [13], EN1994-1-2 [1], EN1992-1-2 [14] and EN1996-1-2 [15]. The arithmetic average temperature of each component is based on the temperature of each node. The 500 °C isothermal is also used to define concrete layer to be neglected, $b_{c,fi,h}$, (see Fig. 4).

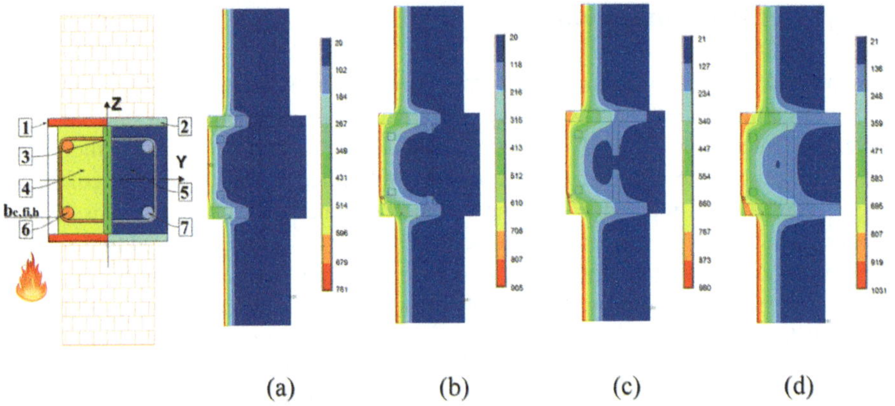

Fig. 4. Identification of the seven components based on the temperature distribution for (a) R30, (b) R60, (c) R90 and (d) R120.

The exposed side is submitted to standard fire conditions, using a convection coefficient of 25 [W/m^{2}K] and an emissivity of the fire equal to 1. The unexposed side is submitted to a convective coefficient of 9 [W/m^{2}K] to include the radiation effect. An initial uniform temperature is applied to all the nodes (20 °C).

The nonlinear transient thermal analysis required an integration time step of 60 s, which can decrease to 0.1 s and increase up to 60 s. The criterion for convergence uses a tolerance value for the heat flow, smaller than 0.1% with a minimum reference value of 1×10^{-6}. The time history for the temperature evolution is depicted in Fig. 5, where one can see the high temperature gradient in direction perpendicular to the web, during 120 min of fire exposure.

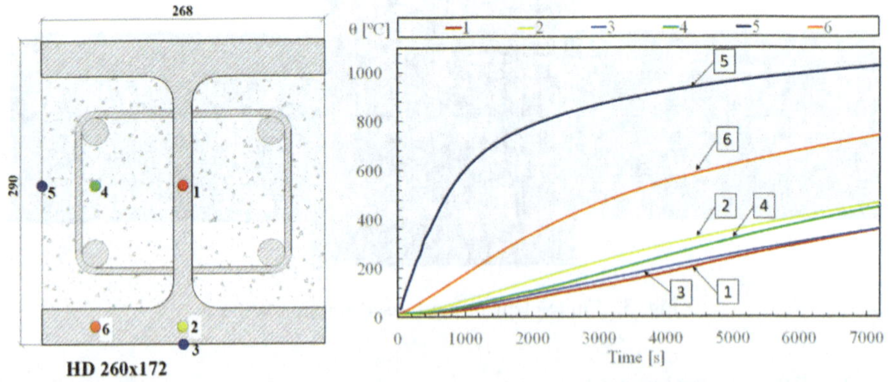

Fig. 5. Temperature evolution in cross section of profile HD 260 × 172.

5 Balanced Summation Model – New Proposal

The new proposal for the balanced summation model requires new analytical formulae to take into consideration the effect of the fire in seven components. The flange, concrete and rebars are divided in two components (exposed to fire and not exposed to fire). The new proposal follows the same principle of the current Eurocode version EN1994-1-2 [1], but they are modified according to Eqs. 8 and 9. The reduction coefficients ϕ, depend on the effect of thermal stresses and were kept almost equal to the current version of EN1994-1-2 [1], with exception to the coefficients $\phi_{f,\theta}$ and $\phi_{s,\theta}$ applied to R120, that should be modified to 0.8.

$$N_{fi,pl,Rd} = N_{fi,pl,Rd,f,e} + N_{fi,pl,Rd,f,ne} + N_{fi,pl,Rd,w} + N_{fi,pl,Rd,c,e} + N_{fi,pl,Rd,c,ne} + N_{fi,pl,Rd,s,e} + N_{fi,pl,Rd,s,ne} \quad (8)$$

$$(EI)_{fi,eff,z} = \phi_{f,\theta}(EI)_{fi,f,e,z} + \phi_{f,\theta}(EI)_{fi,f,en,z} + \phi_{w,\theta}(EI)_{fi,w,z} + \phi_{c,\theta}(EI)_{fi,c,e,z} + \phi_{c,\theta}(EI)_{fi,c,ne,z} + \phi_{s,\theta}(EI)_{fi,s,e,z} + \phi_{s,\theta}(EI)_{fi,s,ne,z} \quad (9)$$

5.1 Flange Components

The flange is divided into two components, half-flange exposed to fire and the half-flange not exposed to fire. New formulae are proposed for the calculation of the average temperature, based on some geometrical factors. Table 3 presents the improved equations for the average temperature of the components, with the respective parameters for each partially encased cross section type. The value $\theta_{fe,t}$ represents the average temperature on the exposed flange and $\theta_{fne,t}$ represents the average temperature on the unexposed flange.

Table 3. New formulae and parameters for average flange temperature.

$\theta_{fe,t} = \theta_{0,t} + k_{fe,f}\,(t_f) + k_{t,fe}\,(A_m/V)$

Standard fire resistance	HEA			HD			UC		
	$k_{f,e}$ [°C/mm]	$k_{t,fe}$ [m°C]	$\theta_{0,t}$ [°C]	$k_{f,e}$ [°C/mm]	$k_{t,fe}$ [m°C]	$\theta_{0,t}$ [°C]	$k_{f,e}$ [°C/mm]	$k_{t,fe}$ [m°C]	$\theta_{0,t}$ [°C]
R30	0.09	36	185	−0.42	45.95	140	−0.55	40.9	150
R60	1.48	45	295	−0.40	59.6	254	−0.53	54.5	262
R90	2.25	54	348	−0.20	69.95	307	−0.34	62.5	326
R120	3.15	58	389	0.01	74	344	−0.20	67.0	375

$\theta_{fne,t} = \theta_{0,t} + k_{f,fne}\,(t_f) + k_{t,fne}\,(A_m/V)$

Standard fire resistance	HEA			HD			UC		
	$k_{f,fne}$ [°C/mm]	$k_{t,fne}$ [m°C]	$\theta_{0,t}$ [°C]	$k_{f,fne}$ [°C/mm]	$k_{t,fne}$ [m°C]	$\theta_{0,t}$ [°C]	$k_{f,fne}$ [°C/mm]	$k_{t,fne}$ [m°C]	$\theta_{0,t}$ [°C]
R30	2.18	22	−66.4	0.11	23.62	−38	0.10	21.8	−31
R60	5.19	41	−139.0	0.30	46.8	−63	0.30	42.0	−45
R90	8.39	65	−245.0	0.72	71.4	−110	0.79	66.2	−96
R120	10.20	85	−305.0	0.92	93.8	−144	0.98	83.8	−110

The mechanical properties should be affected by the average temperature, using the reduction factors presented in EN1994-1-2 [1]. The calculation of the plastic resistance to axial compression and effective flexural stiffness for the flange component are represented in Eq. 10 and Eq. 11, and depend on the geometry and materials properties (yield stress: $f_{ay,f,e,t}$, $f_{ay,f,ne,t}$ and elastic modulus: $E_{a,f.e,t}$, $E_{a,f.ne,t}$).

$$N_{fi,pl,Rd,f} = 2\left[\frac{(b/2\, t_f f_{ay,f,e,t})}{\gamma_{M,fi,a}}\right] + 2\left[\frac{(b/2\, t_f f_{ay,f,ne,t})}{\gamma_{M,fi,a}}\right] \tag{10}$$

$$(EI)_{fi,f,z} = \left[2\left(\frac{\left(t_f (b/2)^3\right)}{12} + \frac{t_f b^3}{32}\right) E_{a,f,e,t}\right] + \left[2\left(\frac{\left(t_f (b/2)^3\right)}{12} + \frac{t_f b^3}{32}\right) E_{a,f,ne,t}\right] \tag{11}$$

5.2 Web Component

The reduction of the web area is not considered in this proposal. Instead a new improvement is proposed, based on the average temperature of the entire web region. The new plastic resistance to axial compression and the effective flexural stiffness in this component use the temperature effect (average web temperature - $\theta_{w,t}$) on the material properties (affecting the yield stress and elastic modulus) for the new parameters (see Table 4).

Table 4. New formulae and parameters for the web average temperature.

$\theta_{w,t} = \theta_{0,w} + k_{w,w}\,(1/t_w) + k_{t,w}\,(A_m/V)$									
Standard fire resistance	HEA			HD			UC		
	$k_{w,w}$ [mm°C]	$k_{t,w}$ [m°C]	$\theta_{0,w}$ [°C]	$k_{w,w}$ [mm°C]	$k_{t,w}$ [m°C]	$\theta_{0,w}$ [°C]	$k_{w,w}$ [mm°C]	$k_{t,w}$ [m°C]	$\theta_{0,w}$ [°C]
R30	90.93	22.00	−50	−251.00	23.80	22.00	−234.62	23.05	20.00
R60	344.12	36.00	−75	−694.57	49.00	30.00	−630.00	50.00	35.00
R90	212.59	64.00	130.	1220.32	79.00	45.00	1142.11	77.49	40.00
R120	578.77	73.00	145	1585.77	102.50	45.00	1500.00	103.00	45.00

The plastic resistance to axial compression and effective flexural stiffness for the web, without any geometry reduction, should be calculated according to Eq. 12. These formulae depend on geometric parameters and materials properties (yield stress $f_{ay,w,t}$ and elastic modulus $E_{a,w,t}$).

$$\begin{aligned} N_{fi,pl,Rd,w} &= \left[e_w (h - 2t_f) f_{ay,w,t}\right] / \gamma_{M,fi,a} \\ (EI)_{fi,w,z} &= \left[E_{a,w,t} (h - 2t_f) e_w^3\right] / 12 \end{aligned} \tag{12}$$

5.3 Concrete Components

The concrete was considered divided into two components, concrete exposed to fire and concrete not exposed to fire. The average temperature of concrete components ($\theta_{ce,t}$ and $\theta_{cne,t}$) requires a new formulae and new parameters according to Table 5. The 500 °C isothermal criterion was used to determine de maximum temperature and the region of concrete to be neglected. The concrete exposed to fire has an horizontal size reduction that should be calculated according to "$b_{ce,fi,h}$", see Table 6. There is no reduction to the unexposed concrete for any fire rating.

Table 5. New formulae and parameters for average concrete temperature.

$\theta_{ce,t} = \theta_{0,t} + k_{ce}(1/t_w) + k_{t,ce}(A_m/V)$									
Standard fire resistance	HEA			HD			UC		
	k_{ce} [°C/mm]	k_t [m°C]	$\theta_{0,t}$ [°C]	kce [°C/mm]	k_t [m°C]	$\theta_{0,t}$ [°C]	k_{ce} [°C/mm]	k_t [m°C]	$\theta_{0,t}$ [°C]
R30	−131	30	24	−37	28.0	33	−172	26.5	50
R60	26	40	39	−155	38.9	70	−444	46.5	60
R90	86	50	35	−449	58.9	75	−785	68.9	50
R120	424	50	55	−764	75.0	85	−943	79.5	60
$\theta_{cne,t} = \theta_{0,t} + k_{f,cne}(1/t_w) + k_{t,cne}(A_m/V)$									
Standard fire resistance	HEA			HD			UC		
	k_{ce} [°C/mm]	k_t [m°C]	$\theta_{0,t}$ [°C]	kce [°C/mm]	k_t [m°C]	$\theta_{0,t}$ [°C]	k_{ce} [°C/mm]	k_t [m°C]	$\theta_{0,t}$ [°C]
R30	−32	10.2	−11.00	−70	10.0	−1	−100	10.50	−3.27
R60	79	22.7	−48.00	−272	27.6	−29	−348	27.80	−29.00
R90	76	39.0	−84.75	−535	47.0	−49	−654	48.00	−51.50
R120	75	53.9	−118.00	−683	56.9	−35	−888	54.80	−17.00

Table 6. New formulae and parameters for horizontal reduction of concrete exposed to fire.

$b_{ce,fi,h} = b_{0,ch} + k_{w,ch}(1/t_w) + k_{t,ch}(A_m/V)$, if $\theta_{c,t} = 500$ °C, then $b_{c,fi,h} = (b - t_w)/2$									
Standard fire resistance	HEA			HD			UC		
	$b_{0,ch}$ [mm]	$k_{t,ch}$ [m²°C]	$k_{w,ch}$ [mm°C]	$b_{0,ch}$ [mm]	k_{tch} [m²°C]	k_{wch} [mm°C]	$b_{0,ch}$ [mm]	k_{tch} [m²°C]	k_{wch} [mm°C]
R30	11.68	0	0	11.68	0.000	0	11.68	0.000	0
R60	25.4	0.05	17	23.25	0.026	−4	23.00	0.026	4.52
R90	28	0.34	−12	31.00	0.180	−12	31.00	0.23	−21
R120	31	0.25	65	38.00	0.400	−26	38.00	0.400	−27

The Eqs. 13 and 14 present the formulae for the calculation of the plastic resistance to axial compression and the calculation of the effective flexural stiffness, for both components (concrete exposed and the concrete not exposed). These formulae depend on geometry and on the concrete materials properties (compressive resistance at elevated temperature, $f_{c,\theta}$ and secant elastic modulus at elevated temperature, $E_{c,sec,\theta}$).

The number 0.86 is a calibration factor. $I_{s,e,z}$ represents the second order moment of area of the reinforcing bars related to the z central axis and $I_{c,e,z}$, $I_{c,ne,z}$ represent the second order moment of the concrete area related to the z central axis, for the exposed and unexposed side, respectively.

$$N_{fi,pl,Rd,c} = \left[\frac{0,86\{((h - 2t_f)(b/2 - t_w/2 - b_{ce,fi,h})/12) - A_s\}f_{c,\theta}}{\gamma_{M,fi,c}}\right] + \left[\frac{0,86\{((h - 2t_f)(b/2 - t_w/2)/12) - A_s\}f_{c,\theta}}{\gamma_{M,fi,c}}\right] \quad (13)$$

$$(EI)_{fi,ce,z} = \left[E_{c,sec,\theta}\left(I_{c,e,z} - I_{s,e,z}\right)\right] + \left[E_{c,sec,\theta}\left(I_{c,ne,z} - I_{s,ne,z}\right)\right] \quad (14)$$

5.4 Rebar Components

The average temperature of reinforcing rebars require a new approximation, using new parameters. The thermal behaviour of the reinforcing bars depends on its geometric position "u" and on the section factor. The rebars are divided in two components, rebars exposed to fire and rebars not exposed to fire. The influence of geometric position "u" is higher for the case of rebars exposed to fire. The new parameters are present in Table 7.

Table 7. New formulae and parameters for the average temperature of exposed rebars.

$\theta_{s,e,t} = \theta_{0,s,e,t} + k_{t,s,e}\,(A_m/V) + k_{u,s,e}(u)$						
Standard fire resistance	HEA			HD/UC		
	$\theta_{0s,e,t}$ [°C]	$k_{t,s,e}$ [m°C]	$k_{u,s,e}$ [m°C]	$\theta_{0s,e,t}$ [°C]	$k_{t,s,e}$ [m°C]	$k_{u,s,e}$ [m°C]
R30	310	0,4	−2,992	140	6	0
R60	640	5,7	−6,719	245	20	0
R90	765	11	−7,400	360	22	0
R120	840	16	−7,600	420	26	0
$\theta_{s,ne,t} = \theta_{0,s,ne,t} + k_{t,s,ne}\,(A_m/V) + k_{u,s,ne}(u)$						
Standard fire resistance	HEA			HD/UC		
	$\theta_{0s,e,t}$ [°C]	$k_{t,s,e}$ [m°C]	$k_{u,s,e}$ [m°C]	$\theta_{0s,e,t}$ [°C]	$k_{t,s,e}$ [m°C]	$k_{u,s,e}$ [m°C]
R30	20	3,5	−0,158	10	4	0
R60	35	12,0	−0,61	−10	16	0
R90	45	17,5	−0,60	−9	25	0
R120	−30	30,0	0,50	6	33	0

The Eqs. 15 and 16 are used for the calculation of the plastic resistance to axial compression and the calculation of the effective flexural stiffness for these components. The reduction of the mechanical properties (elastic modulus E_s and yield stress $f_{s,y}$) is introduced by the average temperature of rebars, using the reduction factors ($K_{s,-,E,\theta}$ and $K_{sy,\theta}$) also presented in EN1994-1-2 [1].

$$N_{fi,pl,Rd,s} = \left[\frac{A_{s,e}k_{sy,\theta}f_{s,y}}{\gamma_{M,fi,s}}\right] + \left[\frac{A_{s,ne}k_{sy,\theta}f_{s,y}}{\gamma_{M,fi,s}}\right] \quad (15)$$

$$(EI)_{fi,s,z} = \left[k_{s,e,E,\theta}E_sI_{s,e,z}\right] + \left[k_{s,ne,E,\theta}E_sI_{s,ne,z}\right] \quad (16)$$

6 Results

New simple formulae are presented, based on the average temperature of each component and based on the residual area of the concrete for different fire ratings (R30, R60, R90 and R120). The comparison of the results is presented in Fig. 6, using the results of the finite element simulation and the results of the new formulae. All the results determined by the new proposal are in the safe side, when the comparison is made with the numerical results.

The plastic resistance to axial compression decreases with time, being the highest reduction verified to R90 and R120. No big reduction is expected to R30. The effective flexural stiffness also decreases with time, being in some cases almost 40% of the effective flexural stiffness at room temperature after 120 min of fire exposure.

The resistance to axial compression and effective flexural stiffness are used for the calculation of columns buckling resistance. The results are calculated through curve "c" of EN 1993-1-1 [12], based on the prescription of the current version of EN 1994-1-2 [1], using the new proposal, for 30, 60, 90 and 60 min of fire exposure, see Fig. 7. These calculations are presented for two different PEC lengths (3 and 4 m) and three different boundary conditions (pinned-pinned, fixed-pinned, and fixed-fixed).

The non-dimensional slenderness depends on the buckling length, under fire conditions, of each partially encased column. The reduction coefficient for the buckling resistance is defined by the curve c of the EN1993-1-1 [12]. The columns with 4 m length present higher non-dimensional slenderness. For that reason, the buckling resistance is smaller. The reduction of the buckling resistance is smaller for the fixed-fixed boundary condition, in comparison to the other boundary conditions. A few columns with 3 m length, after 30 min of fire exposure, are not susceptible to prone into buckling.

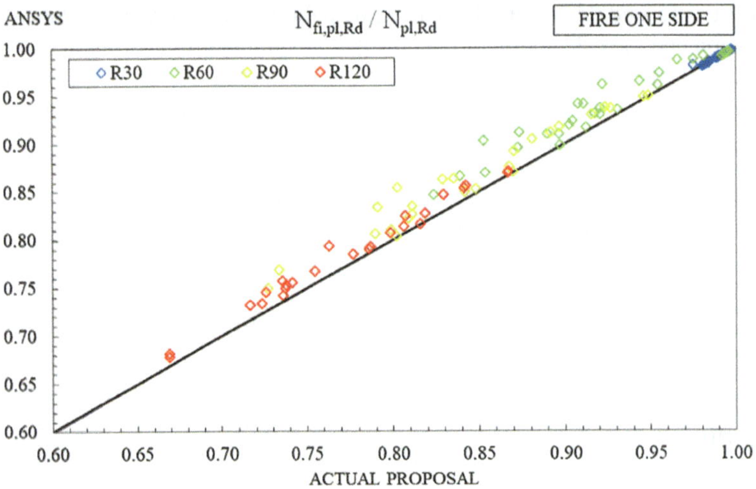

(a)

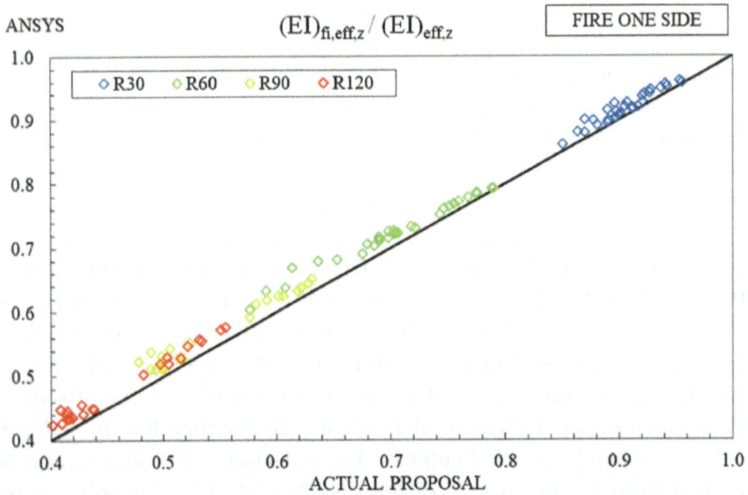

(b)

Fig. 6. Comparison results between the numerical simulation and improved formulae for the calculations of the resistance to axial compression (a) and effective flexural stiffness around the weak axis (b).

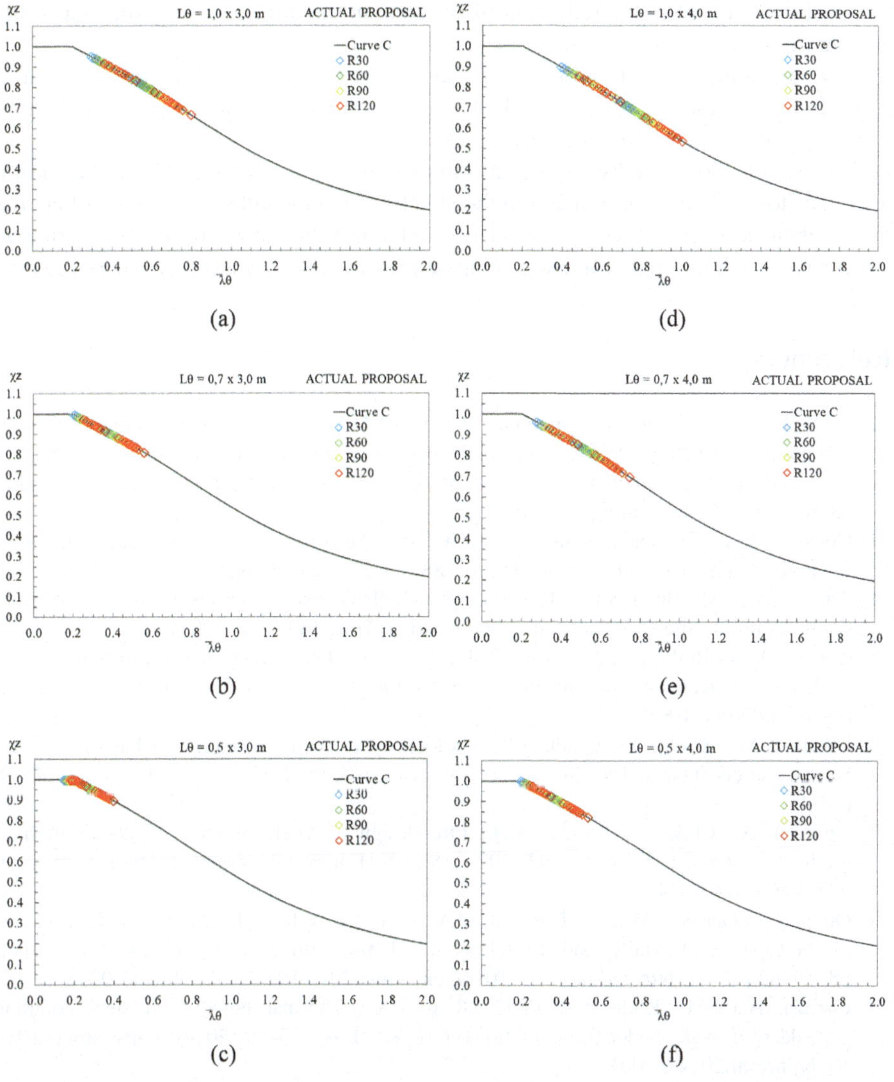

Fig. 7. Buckling curve for the lengths L = 3 m (a, b, c) and L = 4 m (d, e, f) for the three buckling lengths, for scenario fire one side, according to the Actual Proposal.

7 Conclusion

The simplified calculation method, proposed in annex G of EN1994-1-2 [1], does not provide any formulae for the fire design of PEC embedded in walls. The new proposal, presented herein, consider the PEC embedded in wall for the specific fire scenario (fire from one side). The wall partially protects the composite cross section from the fire effect, avoiding the reduction of the concrete region in the direction of the web. This proposal only considers the reduction of concrete in the direction perpendicular to the web and only in the exposed side block.

The maximum reduction to the plastic resistance to axial compression, after 120 min of fire exposure is 72%, while the maximum reduction to the effective flexural stiffness is 40% for the same fire resistance.

The results obtained by numerical simulation for the calculation of the plastic resistance to axial compression and to the effective flexural stiffness are bigger than the results obtained from this new proposal, due to the fact that the approximation formulae are always providing higher average temperatures for each component (safety reason).

References

1. CEN- European Committee for Standardization, EN 1994-1-2: Design of composite steel and concrete structures. Part 1-2: General rules - Structural fire design, Brussels (2005)
2. ISO834-1, Fire-resistance tests - Elements of building construction - Part 1: General requirements, Switzerland, p. 25 (1999)
3. Cooke, G.M.E.: Thermal bowing and how it affects the design of fire separating construction. In: Proceedings of Interflam, London, vol. 88, pp. 230–236 (1988)
4. Garlock, M.E.M., Quiel, S.E.: Mechanics of wide-flanged steel sections that develop thermal gradients due to fire exposure. Int. J. Steel Struct. **7**(3), 153–162 (2007)
5. Correia, A.M., Rodrigues, J.P., Silva, V.P.: A simplified calculation method for temperature evaluation of steel columns embedded in walls. Fire Mater. **35**, 431–441 (2011). https://doi.org/10.1002/fam.1063
6. Dwaikat, M., Kodur, V., Quiel, S.E., Garlock, M.E.M.: Experimental behavior of steel, beam-columns subjected to fire-induced thermal gradients. J. Constr. Steel Res. **67**, 30–38 (2011)
7. Agarwal, A., Choe, L., Varma, A.H.: Fire design of steel columns: effects of thermal gradients. J. Constr. Steel Res. **93**, 107–118 (2014). https://doi.org/10.1016/j.jcsr.2013.10.023. ISSN 0143-974X
8. Quiel, S.E., Garlock, M.E.M., Dwaikat, M.M.S., Kodur, V.K.R.: Predicting the demand and plastic capacity of axially loaded steel beam–columns with thermal gradients. Eng. Struct. **58**, 49–62 (2014). https://doi.org/10.1016/j.engstruct.2013.10.005. ISSN 0141-0296
9. Correia, A.J.P.M., Rodrigues, J.P.C., Real, P.V.: Thermal bowing on steel columns embedded on walls under fire conditions. Fire Saf. J. **67**, 53–69 (2014). https://doi.org/10.1016/j.firesaf.2014.05.001
10. Ojeda, O.D., Maljaars, J., Abspoel, R.: Fire exposed steel columns with a thermal gradient over the cross-section. Thin-Walled Struct. **98**, 103–110 (2016). https://doi.org/10.1016/j.tws.2015.02.009
11. Rocha, F.M., Rodrigues, J.P.C., Munaiar Neto, J.: Fire behavior of steel and partially encased composite columns embedded on walls. J. Constr. Steel Res. **149**, 105–118 (2018). https://doi.org/10.1016/j.jcsr.2018.07.014
12. CEN- European Committee for Standardization, Eurocode 3: Design of steel structures - Part 1-1: General rules and rules for buildings, Brussels (2005)

13. CEN- European Committee for Standardization, EN 1993-1-2: Design of steel structures - Part 1-2: General rules - Structural fire design, Brussels (2005)
14. CEN- European Committee for Standardization, EN 1992-1-2: Design of concrete structures - Part 1-2: General rules - Structural fire design, Brussels (2004)
15. CEN - European Committee for Standardization, Eurocode 6: Design of masonry structures - Part 1-2: General rules - Structural fire design, Brussels (2005)

Numerical Analysis of Cellular Steel Beams Failure Modes in Fire Conditions

Jaqueline Silva[1], Paola Dalcanal[2], and Luís Mesquita[3(✉)]

[1] Instituto Politécnico de Bragança, Campus de Santa Apolónia, 5301-857 Bragança, Portugal
jaqueline.as@outlook.com

[2] Departamento Acadêmico de Engenharia Civil, Universidade Tecnológica Federal do Paraná (UTFPR), câmpus de Pato, Branco, Brazil
paolardalcanal@utfpr.edu.br

[3] Institute for Sustainability and Innovation in Structural Engineering (ISISE), Instituto Politécnico de Bragança, Campus de Santa Apolónia, 5301-857 Bragança, Portugal
lmesquita@ipb.pt

Abstract. Cellular Beams with long spans are increasingly used in the steel building constructions and usually are used with openings for technical equipment and ventilations systems cut in their web. These elements are deeper than normal rolled sections, with distributed circular openings, or other like squared or hexagonal, produced by cutting and welding hot rolled steel sections.

Cellular beams in fire conditions may collapse by local buckling instability of the web-post between the openings, lateral torsional buckling or by Vierendeel bending at the openings.

This behavior is analyzed and evaluated by a numerical method, using the finite element method, considering geometrically and material nonlinear simulations. A parametric analysis is performed for different cellular beams, considering the variation of: (i) different cross sections; (ii) web post widths; (iii) hole diameter to the section height ratios and (iv) distance between holes. Additionally the residual stress influence in the moment resistance is analyzed considering different distributions.

The cellular beams collapse modes are related to the length of the beam rather to the geometric parameters of the apertures. Short beams, with relatively small T height, collapse by Vierendeel mechanism, for higher heights the collapse is due to web-post buckling. Longer beams collapse mainly by lateral torsional buckling. The Lateral Torsional Buckling resistance determined by the Eurocode 3 part 1.1 formulae, and considering the 2T section, is higher than the resistance obtained by the numerical method. For the studied cases, the presence or absence of residual stresses in the web does not influence the resistant capacity or collapse mode.

Keywords: Cellular beams · Fire conditions · Numerical analysis · Global buckling · Local buckling · Lateral torsional buckling

P. A. G. Piloto et al. (Eds.): CILASCI 2019, LNCE 1, pp. 78–92, 2020.
https://doi.org/10.1007/978-3-030-36240-9_6

1 Introduction

Cellular beam generally originates from steel hot-rolled I-section member, which the web is cut, displaced and subsequently welded (Westok method) to obtain openings in the web spaced regularly, with different forms like circular, sinusoidal or hexagonal, see Fig. 1. The procedure results in a deeper beam, approximately 40–60% deeper than its parent solid section [1]. This type of beam is increasingly used because it is capable to support long spans, it can have a bending resistance up to 2.5 times higher than its parent section and its openings allow the passage of technical equipment and ventilation systems through their web, which reduces the floor height and therefore the overall height of building.

The presence of the opening influences in the failure modes of the beam. Vierendeel mechanism and web post buckling (WPB) are examples of failure modes originate due the openings, while already existing failure modes on solid beams are modified in cellular beams, such as lateral-torsional buckling (LTB) [2]. Besides that, exposure to high temperatures has a detrimental effect on the steel resistance, which added to the initial imperfections of the beam results in the combination of failure modes previously mentioned.

For that reason, the main purpose of the study is to define the failure modes, in particular WPB and LTB, and the beam resistance, relating the temperature with variant geometric parameters, comparing the results obtained by simple calculation methods proposed by Eurocode 3 [3], and Steel Construction Institute (SCI) [4] with numerical results. A finite element model is developed in the software Ansys to study the influence of different geometric parameters, considering geometric imperfections and residual stresses in the beams.

Fig. 1. The cutting process of cellular beams. The use of cellular beams in buildings.

2 Simple Calculation Methods

In the present paper, the parent solid section of the cellular beams is a hot rolled I-section IPE500 of class 1 cross-section. It was chosen the steel grade S355 to the study, because of it is the most applied in cellular beams construction due its greater resistant capacity. The cellular beam is considered a double-symmetric beam with the main geometric dimensions presented in Fig. 2, such as the variation of spacing between the web openings (S) and the web-post width (S_0), with the length of the member (L) varying in each numerical simulation.

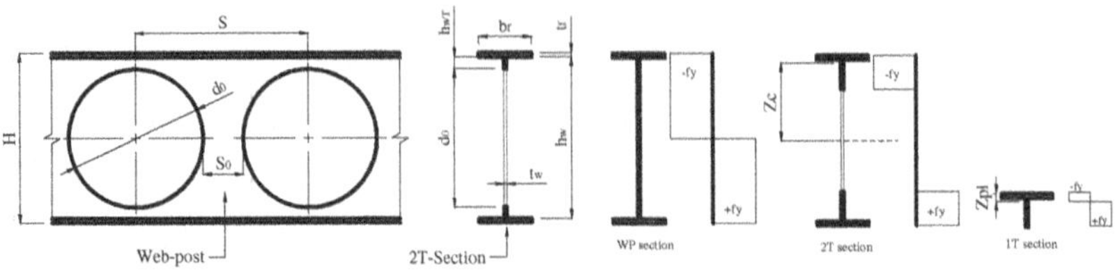

Fig. 2. Cellular beam geometric parameters and plastic stress distributions.

Besides the already mentioned parameters, the other dimensions of the beam illustrated are opening diameter (d_0), flange width (b_f), flange thickness (t_f), height of one Tee (h_{wT}), height of the web (h_w) and total beam height (H). Some geometric parameters vary in relation to the height of the parent section (h) or the opening diameter according to their use, see Table 1. Thus, this paper considers these variations during the numerical simulations, according to the specific focus shown in Table 2.

In a cellular beam submitted to a design degree of utilization in fire conditions, web post failure may occur before the section reaches the critical temperature obtained for members in bending. This fact is mainly due to web height, slenderness and because the temperature of the web-post increases at a faster rate compared to its equivalent (similar web size) solid beam. Beam failure may occur by local buckling instability of the web-post or by Vierendeel bending at the openings.

Table 1. Variation of dimensions according to the application [5].

Objective	Application	d_0	S	H
Optimization of the height/weight ratio	Roofing, Footbridges and Wide-Span purlins	1.0–1.3 h	1.1–1.3d_0	1.4–1.6 h
Optimization of load/weight ratio	Floors, Parking structures and Offshore structures	0.8–1.1 h	1.2–1.7d_0	13–1.4 h

Table 2. Cases analyzed according to the specific focus.

Focus	Type	d_0	S	H	L (m)	T (°C)
Elastic LTB	Solid	–	–	h	2–10	20
	Cellular	0.8 h, 1.0 h, 1.3 h	1.4d_0, 1.7d_0	1.3 h, 1.5 h, 1.7 h	2–10	20
Comparison between solid and cellular beams	Solid	–	–	h	2–10	20, 500, 600, 700
	Cellular	1.0 h	1.4d_0	1.5 h	2–10	20, 500, 600, 700
Spacing between web openings effect	Cellular	1.0 h	1.4d_0	1.5 h	3, 5, 7, 9, 10	20, 500, 700
	Cellular	1.0 h	1.7d_0	1.5 h	3, 5, 7, 9, 10	20, 500, 700

The only design guides for beams with web openings at ambient temperature are the annex N from Eurocode EN1993-1-1 [6], draftEN1993-1-13 [7], P100 [8] and P355 from SCI [4], and no guides are available for these beams at elevated temperatures. Considering this references and the Eurocode for fire design EN1993-1-2 [9], the formulae for the different ultimate limit states are presented. These rules follow, the material properties temperature variation accordingly to EN1993-1-2 [9].

2.1 Plastic Bending Resistance

The plastic bending resistance of the beam is calculated by its geometric dimensions and its material yield stress, for solid and cellular beams. Thus, the expressions based on [3, 4, 9] are:

$$M_{fi,\theta,Rd} = k_{y,\theta}\left[\gamma_{M,0}/\gamma_{M,fi}\right]M_{c,Rd} \tag{1}$$

$$M_{c,Rd,s} = W_y f_y/\gamma_{M0} \tag{2}$$

$$M_{c,Rd,2T} = 2A_T Z_c f_y/\gamma_{M0} \tag{3}$$

$$M_{pl,Rd,1T} = \frac{A_{wT}f_y}{\gamma_{M0}}\left(0,5h_{wT} + t_f - Z_{pl}\right) + \frac{A_f f_y}{\gamma_{M0}}\left(0,5t_f + Z_{pl} - \frac{Z_{pl}^2}{t_f}\right) \tag{4}$$

$$M_{wp,Rd} = S_0^2 t_w f_y/6\gamma_{M0} \tag{5}$$

Where $M_{fi,\theta,Rd}$ is the design moment resistance for a uniform temperature, $k_{y,\theta}$ is the reduction factor for the yield stress at the temperature θ, $M_{c,Rd,s}$ is the plastic bending resistance of the solid beam, $M_{c,Rd,2T}$ is the plastic bending resistance of two Tee, $M_{pl,Rd,1T}$ is the plastic bending resistance of one Tee, $M_{wp,Rd}$ is the bending resistance of a web-post, W_y is the plastic section modulus about y-axis ($W_y = W_{pl,y}$ for class 1 or 2 cross-sections), f_y is the yield stress of steel, A_T is the area of one Tee, Z_c is the distance from the local centroid of the area of one Tee to the global centroid, A_{wT} is the area of the web of one Tee, A_f is the area of the flange of one Tee, Z_{pl} is the depth of plastic neutral axis of Tee from outer face of flange = $(A_f + A_{wT})/(2b_f)$, $\gamma_{M,fi}$ and γ_{M0} are partial safety factors in fire situation and ambient temperature, respectively, $\gamma_{M0} = \gamma_{M,fi} = 1$.

2.2 Lateral-Torsional Buckling Resistance

The calculation of the lateral-torsional buckling resistance ($M_{b,Rd}$) at ambient temperature (20 °C) of a laterally unrestrained steel beam is done by the following expressions [3]. The difference between the calculation of solid and cellular beam is the value adopted to some constants, like warping constant, torsional constant and moment of inertia, which vary with the geometric parameters adopted, hence it must be calculated for each beam. In accordance with the 2T-approach, cross-sectional properties are computed at the center of the web opening. The lateral torsional buckling resistance in fire is calculated according to the Eurocode 3 part 1-2 [9], formulae, presented in next equations.

$$M_{b,fi,t,Rd} = \chi_{LT,fi} W_{pl,y} k_{y,\theta,com} f_y / \gamma_{M,fi} \tag{6}$$

$$\chi_{LT,fi} = \frac{1}{\phi_{LT,\theta,com} + \sqrt{\left[\phi_{LT,\theta,com}\right]^2 - \left[\bar{\lambda}_{LT,\theta,com}\right]^2}} \tag{7}$$

$$\phi_{LT,\theta,com} = 0,5\left[1 + \alpha\bar{\lambda}_{LT,\theta,com} + \left(\bar{\lambda}_{LT,\theta,com}\right)^2\right] \tag{8}$$

$$\alpha = 0,65\sqrt{235/f_y} \tag{9}$$

$$\bar{\lambda}_{LT,\theta,com} = \bar{\lambda}_{LT}\left[k_{y,\theta,com}/k_{E,\theta,com}\right]^{0,5} \tag{10}$$

Where $M_{b,fi,t,Rd}$ is the design buckling resistance moment, $\chi_{LT,fi}$ is the reduction factor for LTB in fire situation, $k_{y,\theta,com}$ is the reduction factor for the yield strength and $k_{E,\theta,com}$ is the reduction factor for the slope of the linear elastic range, both at the maximum temperature in compression flange, $\phi_{LT,com}$ value to determine $\chi_{LT,fi}$, $\bar{\lambda}_{LT,\theta,com}$ and $\bar{\lambda}_{LT}$ are the non-dimensional slenderness for LTB at temperature θ and ambient temperature, respectively, α is the imperfection factor. The non-dimensional slenderness at ambient temperature is calculated considering the 2T plastic moment and elastic critical moment.

In the study of Real et al. [10], it is suggested to change the α and $\chi_{LT,fi}$ calculation for solid beams at high temperatures, α varying according to the ratio between beam height and width, type of cross-section and steel grade, and $\chi_{LT,fi}$ according to the type of moment distribution.

$$\alpha = \beta\sqrt{235/f_y} \tag{11}$$

$$\chi_{LT,fi,mod} = \chi_{LT,fi}/f \tag{12}$$

$$f = 1 - 0.5(1 - k_c) \tag{13}$$

Where β is the severity factor that is equal to 0.75 for rolled I-section with steel grade S355 and h/b > 2, $\chi_{LT,fi,mod}$ is a modified reduction factor, f is a value to determine $\chi_{LT,fi,mod}$ and k_c is a correction factor that is equal to 0.91 when the distribution moment is parabolic.

2.3 Shear Resistance

The shear resistance is determined using the shear area of the section (A_V), which for solid beam corresponds to the area of the web and for cellular beam is computed as [4]:

$$A_V = A_T - b_f t_f + (2r + t_w)0,5 t_f \tag{14}$$

Where r is the fillet radius of the parent section. Therefore, the shear resistance of the 2T-section ($V_{pl,Rd,2T}$) and the shear resistance for web-post section ($V_{wp,Rd}$) are given by

$$V_{pl,Rd,2T} = \frac{2A_V f_y}{\gamma_{M0}\sqrt{3}} \quad (15)$$

$$V_{wp,Rd} = \frac{t_w S_0 f_y}{\gamma_{M0}\sqrt{3}} \quad (16)$$

2.4 Web-Post Buckling Resistance

This local failure mode occurs only in cellular beams due to the combination of horizontal shear and double curvature bending across its height, which causes an inclined compression line across the web post height [2]. The effect of the presence of an aperture in the web is associated with one inclined edge of the opening is stressed in tension and the opposite edge in compression, a twisting action along the height of the web-post is caused by the buckling [11], see Fig. 3.

That resistance depends on the web-post width (S_0), i.e. the spacing between the edges of the openings, which is considered a beam with closely spaced openings when the $S_0 \leq d_0$ and largely spaced openings when $S_0 > d_0$ [4]. The width influences directly the effective web length (l_e), the length subjected to buckling. The study focuses only on beams with closely spaced openings.

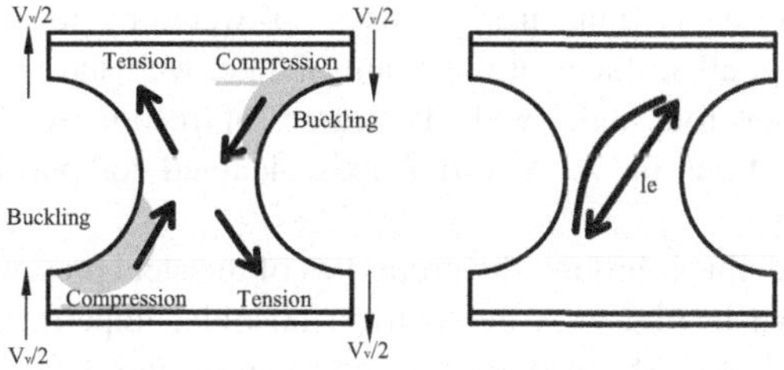

Fig. 3. Strut model of web-post buckling.

The buckling resistance of the web post ($N_{wp,Rd}$) is determined by the following expressions [4].

$$N_{wp,Rd} = \chi \frac{S_0 t_w f_y}{\gamma_{M1}} \quad (17)$$

$$\chi = \frac{1}{\phi + \sqrt{\phi^2 - \bar{\lambda}^2}} \text{ but } \chi \leq 1,0 \quad (18)$$

$$\phi = 0,5\left[1 + \alpha(\bar{\lambda} - 0,2) + \bar{\lambda}^2\right] \quad (19)$$

$$\bar{\lambda} = \frac{2,5\sqrt{S_0^2 + d_0^2}}{t_w}\frac{1}{\lambda_1} \quad (20)$$

Where χ is the reduction factor, α is the partial factor according to the buckling curve, it is chosen the buckling curve b for cellular beams from rolled sections [4], $\bar{\lambda}$ is the non-dimensional slenderness of the web-post, λ_1 the geometric slenderness, calculated as in EN1993-1-1 [3] (Fig. 4).

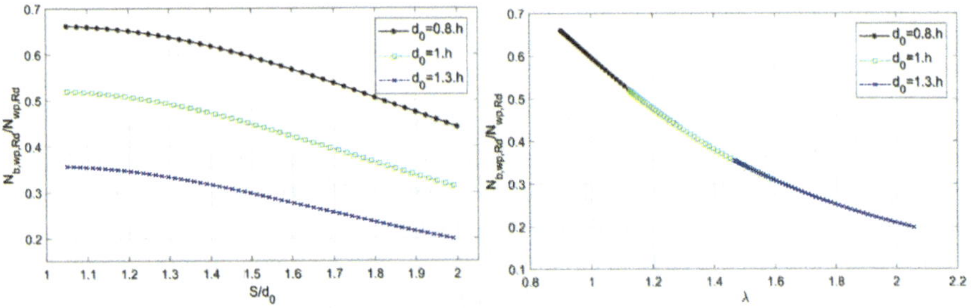

Fig. 4. Web-post buckling resistance in function of the geometric dimensions.

3 Finite Element Numerical Model of Solid and Cellular Beams

The numerical analysis is done using Ansys software, by means of the structural element Shell 181 on all surfaces of the beam (flange, web and reinforcing end plate). This finite element has four nodes with six degrees of freedom in each node, which are translations and rotations on X, Y and Z axis, adapted for plasticity and large displacement behavior.

The material is defined through the properties of the steel presented in Eurocode EN 1993 1-2 [9], shown in Fig. 5. A geometric sinusoidal imperfection is added to the beam with amplitude L/1000 in mid-span, considering the deformed shape from the first buckling mode resulting from the elastic Eigen Buckling analysis. Despite this, Sect. 4 also presents and compares results without geometric imperfection.

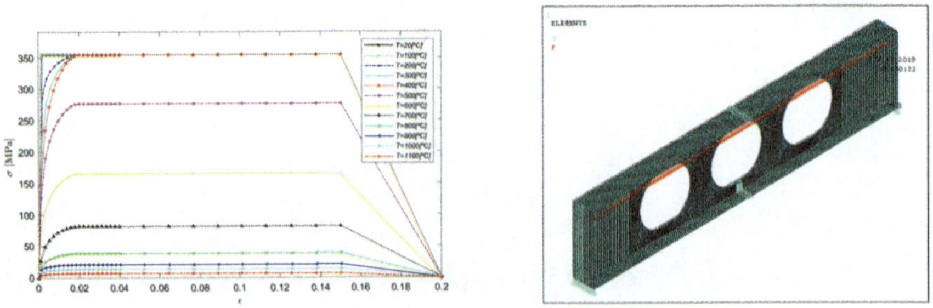

a) b)

Fig. 5. (a) Stress vs Strain relationship at elevated temperatures. (b) Finite element mesh for a cellular beam with: L = 3 m, d0 = 1.0 h, H = 1.3 h.

In all the studied cases the beams are considered simply supported, with the bottom flange at the ends restrained in vertical direction, the web at the ends restrained in lateral direction, and due the symmetry all nodes at mid-span restrained axially, as presented in Fig. 5. The mechanical load is applied as a distributed load in all the centre nodes from the top flange. An iterative procedure was implemented where the load is varying in the simulation time with a minimum time increment corresponding to a minimum nodal load of 0.01 N. The Newton Raphson method was used with a force and displacement nonlinear convergence criteria tolerance equal to 0.001.

The model also considers the influence of residual stress along the beam and its intensity is defined according to the yield stress (f_y = 355 MPa). For solid beams, the residual stress follows the bi-triangular distribution with maximum amplitude equal to $0.3f_y$, as shown in Fig. 6(a).

For the cellular beams, the residual stress distribution changes, mainly due to the cutting and welding procedure. The introduction of the openings mostly changes the residual stress in the web, due to the insertion of heat in that area. So, the residual stress in the flange is considered equal to solid beams and in the web follows the model proposed in the ECCS document [12], for a plate with one cut edge, see Fig. 6(b) and the detail (c). It is adopted a tensile stress equal to the yield stress in a narrow strip next to the cut (σ_r), with the variable width equal to "c", the opposite edge with the stress value equal to the value considered for solid beams, $\sigma_t = 0.3f_y$, and the compression stress (σ_c) calculated with Eq. (21), in order to achieve static equilibrium.

$$\sigma_c = \sigma_r \frac{c(4h_{wT} - c)}{(b - c)^2} \tag{21}$$

A more simplified residual stress pattern follows the recommendation of ECCS (1984) for solid beams [13], which adopts the triangular distribution from Fig. 6(a) for the flange, while the stresses at the web are considered zero, as presented in Fig. 6(d). This assumes that the whole cutting process dissipates the constraint on the web. This model is used by different authors with the assumption that the lateral torsional buckling is mainly affected by residual stresses applied in flanges [14, 15].

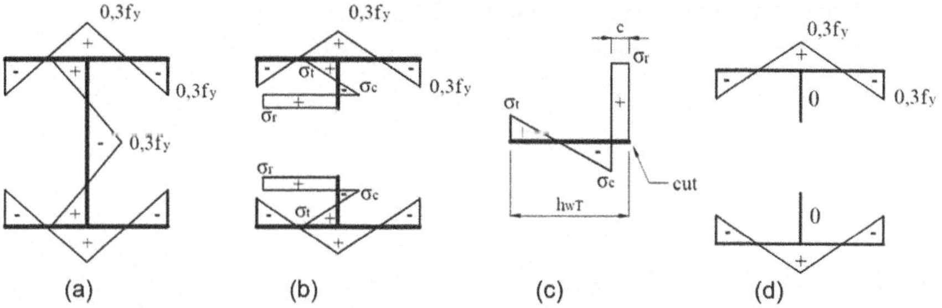

Fig. 6. (a) Residual stresses on solid beams. (b) Residual stresses on cellular beams (TR2). (c) Stress distribution in a plate with one cut edge. (d) Simplified residual stress model (TR1).

4 Comparison of Simplified and Numerical Results

The Eigen buckling analysis were performed for all solid and cellular beams studied. Figure 7 presents these results for solid and cellular beams of different geometric dimensions (marks). These numerical results are compared with the analytical equation from the ENV1993-1-1 [6], represented by the dotted lines, in which all geometric properties consider the 2T cross section dimensions.

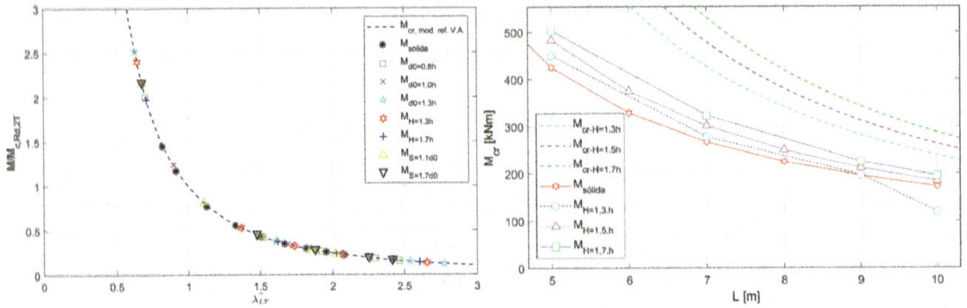

Fig. 7. Lateral torsional buckling elastic critical moment for solid and cellular beams.

The figure shows that for intermediate and long span cellular beams the Ansys critical moment is smaller than the ones obtained by the Eurocode 3 formulae, using the 2T section approach.

An iterative procedure was implemented using material and geometric non-linear behavior and mechanical load increments along the time, considering a uniform temperature distribution over the whole beams. All nonlinear simulations were performed until the beams collapse, defined as the time/load when the numerical convergence was not attained.

Figure 8 presents the numerical simulations of solid beams at ambient and elevated temperatures and their comparison with the simplified methods. In this figure the numerical results consider the Ansys cross section plastic resistance and beam elastic critical moment. The Eurocode 3 part 1-2 [9], formulae and the one improved by Vila Real [10], are compared for fire situation.

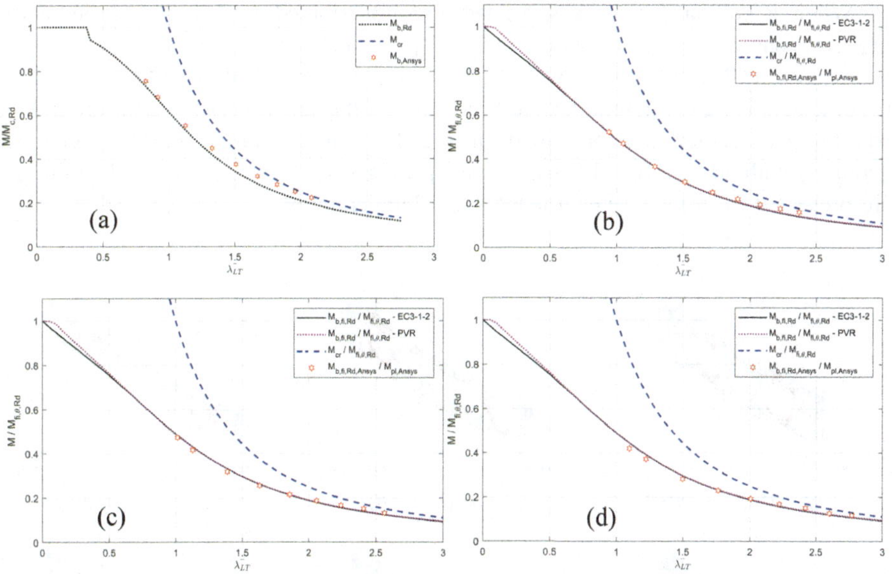

Fig. 8. Lateral torsional Buckling resistance of solid beams at different temperatures (a) 20 °C (b) 500 °C (c) 600 °C (d) 700 °C.

A comparison study was done to verify the influence of the residual stress distribution for cellular beams. Several beam lengths were considered for the geometric section with high H = 1.5 h, whole diameter equal to h, and whole spacing's S = 1.4d0. The results are presented in Fig. 9. It can be seen that for ambient temperature and at 500 °C the results with TR1 and TR2 give approximately the same results for LTB resistance. This is in line with the knowledge that LTB is more influenced by the flange residual stresses. From this simulations, and due to numerical problems verified in the Newton Raphson method convergence at elevated temperatures, mainly at 800 °C, the next simulations consider the residual stress distribution from TR1.

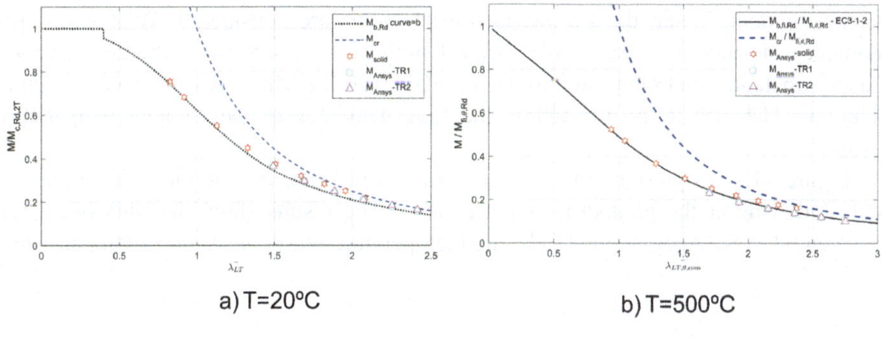

Fig. 9. Influence of the residual stress distribution on the LTB resistance.

When a cellular beam is subjected to a distributed load, and the geometric imperfections are not considered, the collapse mechanisms are due to shear and bending of the T section or the web post, or even by a combination of both promoting in some geometric configurations the Vierendeel collapse with plastic joints around the wholes. This modes are presented in Fig. 10, where, considering the reference geometry with d0 = h, H = 1.5 h and with spacing's S = 1.4d0, the stress distributions are presented for different beam lengths and temperatures.

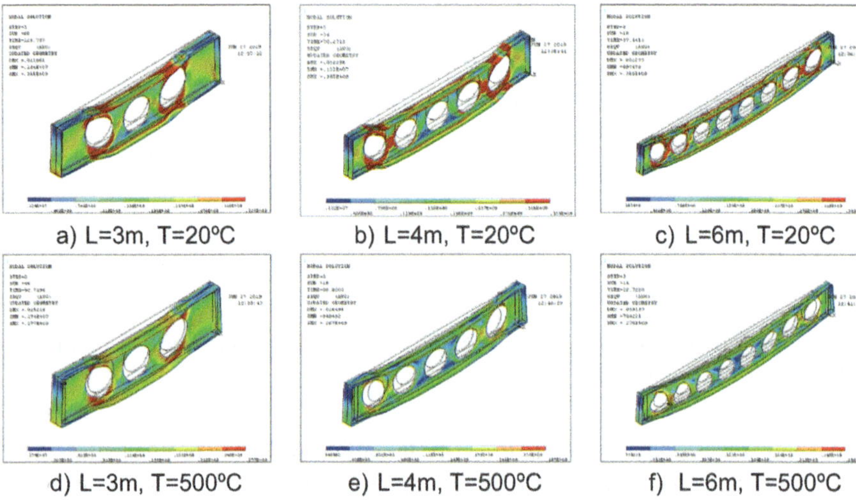

Fig. 10. Collapse modes of cellular beams without geometric imperfections at ambient temperature and at T = 500 °C.

For the same cellular beam geometry, Fig. 11 present the applied collapse maximum moment normalized by the analytical 2T cross section plastic moment resistance. The results at ambient temperature are compared to the results at T = 500 °C for different beam lengths, without considering the geometric imperfections (NGI) and considering it (WGI). It can be seen that for the NGI cases, only higher spans fail by bending, while for shorter beams the failure occurs for loads or moments below the 2T plastic moment. When the geometric imperfections are considered, WGI cases, the moment resistance increases until a beam length of 4 m, defined by the cross section plastic resistance, and then starts to decrease as a result of the lateral torsional buckling behavior. The 500 °C results follow the same tendency of the ambient temperature results.

Figure 11, also presents the resistance of a 6 m beam in function of temperature, with and without the geometric imperfections. The results show that this beam fail between 60 to 80% below the 2T elevated temperature plastic resistance moment, when

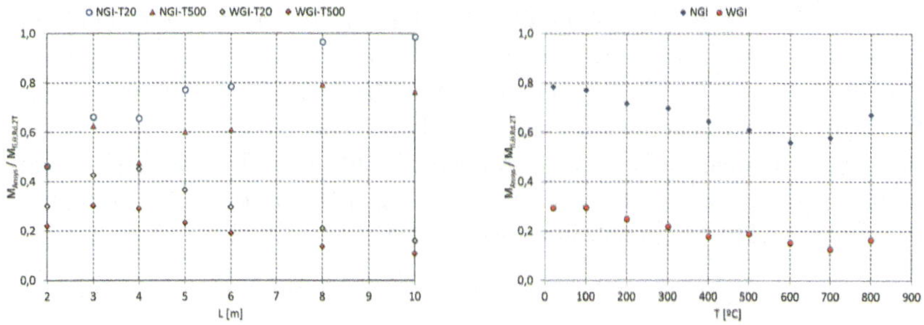

Fig. 11. Cellular beams collapse resistance with and without geometric imperfections in function of the temperature.

is laterally restrained. When LTB collapse mode occur, the resistance reduces to values between 17 to 30%, for temperatures ranges until 800 °C.

In cellular beams, the collapse mode varies with the geometric dimensions. Considering the distance between wholes equal to S = $1.4d_0$, and for beam lengths of 2 and 3 m, the collapse mode was due to web-post buckling, for all the other analysed was due to LTB. Figure 12 shows the beams von Mises stress distribution in the deformed state at the collapse time/load.

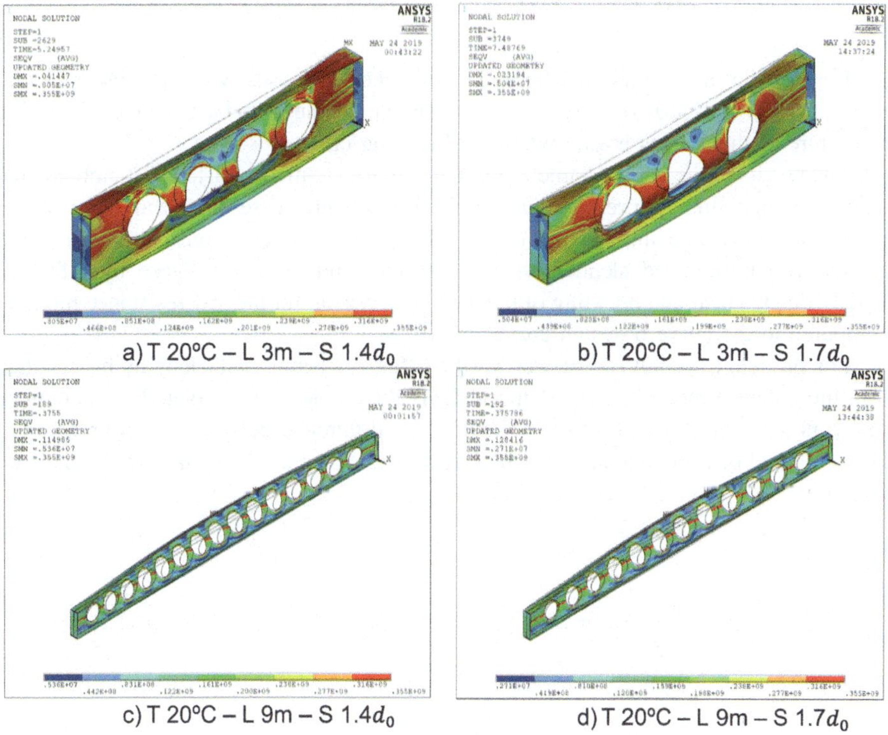

Fig. 12. Collapse mode of cellular beams at ambient temperature.

For the cases in which the first buckling mode is not due to lateral torsional buckling, but local buckling or a combination of local and distortional buckling, as for short cellular beams and high section cellular beams, a perturbation method was applied with a lateral load to promote the LTB collapse mode.

A comparison study was made for different lengths and distances between wholes. These results are presented in the next figure (Fig. 13) in comparison to the buckling curve b for the LTB design and the elastic curve.

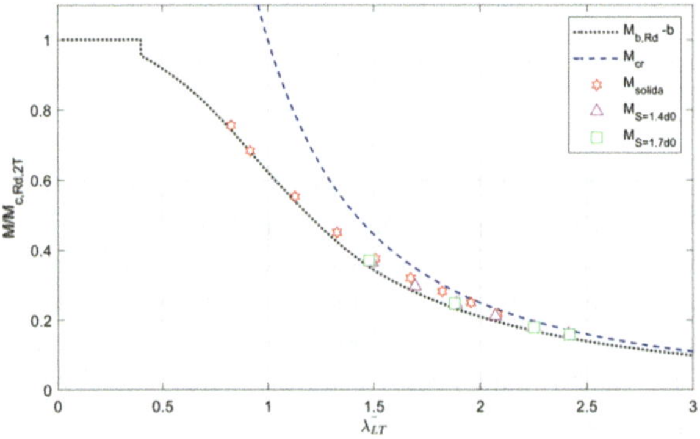

Fig. 13. Resistance moment of cellular beams at ambient temperature.

Besides the lateral torsional buckling resistance increases with the spacing increment, from S = 1.4d0 to 1.7d0, the normalization results indicates that the design procedure of this beams is safe when the buckling curve b is used.

The results of the whole diameter influence in the moment resistance at ambient and elevated temperatures is presented in Fig. 14. For the three cases analyzed, the diameter is related with h and the spacing to the whole diameter ratio constant. The cellular beams with normalized slenderness smaller than unit do not collapse by LTB but mainly due to shear and buckling of the web post, reason for the low resistance moment compared to the curve b design curve.

For the lengths analyzed, it was verified that for beams with lengths higher than 4 m the collapse mode is always due to LTB for ambient and elevated temperatures. For temperatures between 500 and 700 °C the difference between the simple calculation method of Eurocode and the numerical results increases with the whole diameter increase.

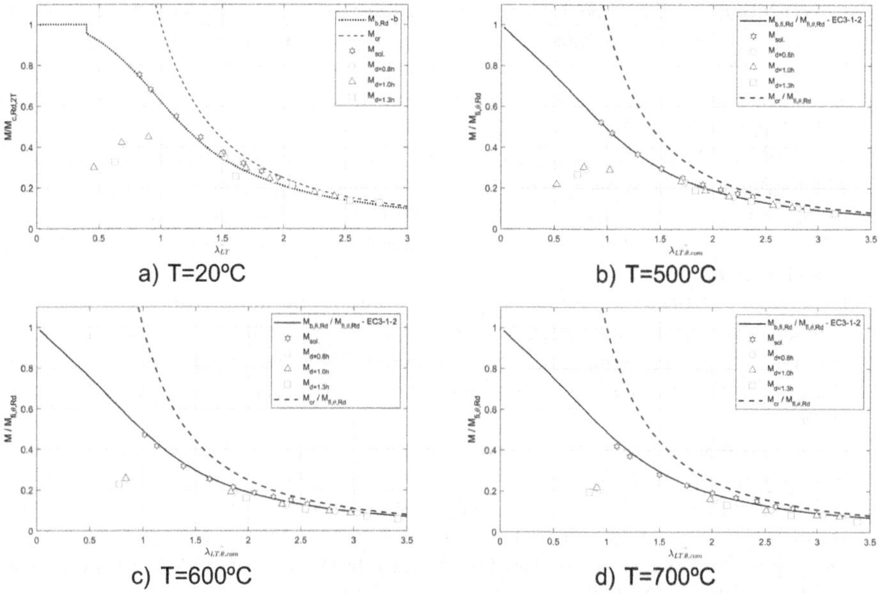

Fig. 14. Influence of the whole diameter in the moment resistance at ambient and elevated temperatures.

5 Conclusions

This work presented a simplified method based on the Eurocode formulae and in the SCI reference for the design of cellular beams at ambient and elevated temperatures.

The structural behavior of solid and cellular beams were compared using nonlinear geometric and material finite element simulations at different temperatures. Cellular beams resistance varies with the most relevant geometric parameters, namely, the section high, whole diameter and whole spacing.

Even though a cellular beam can have a bending resistance up to 2.5 times higher than its parent solid section new collapse modes, compared to solid sections, can reduce the overall beam resistance, especially for low span beams.

For low span cellular beams, and for the geometric dimensions studied, the achieved collapse modes are due to web post shear and buckling and Vierendeel plastic collapse. For longer spans the collapse mode is mainly due to lateral torsional buckling.

References

1. Pachpor, P.D., et al.: Finite element analysis and comparison of castellated and cellular beam. Adv. Mater. Res. **264–265**, 694–699 (2011)
2. Sonck, D.: Global buckling of castellated and cellular steel beams and columns. In: Department of Structural Engineering, Faculty of Engineering and Architecture - Ghent University, p. 349 (2014)

3. CEN: EN1993-1-1 Eurocode 3: Design of steel structures - Part 1-1: General rules and rules for buildings. European Committee for Standardization: Brussels, Belgium (2005)
4. Lawson, R.M., Hicks, S.J.: Design of composite beams with large web openings: in accordance with Eurocodes and the UK National Annexes (P355), ed. SCI. 2011, Silwood Park, Ascot, Berkshire SL5 7QN UK. SCI
5. S.A., A.C.S.: ACB® and Angelina® beams - A new generation of castellated beams, ArcelorMittal, Editor, p. 64 (2018)
6. CEN: ENV 1993-1-1:1992/A2 Eurocode 3: Design of steel structures - Part 1-1: General rules and rules for buildings, Annex N. European Committee for Standardization: Brussels, Belgium (1998)
7. CEN, Draft EN 1993-1-13, Eurocode 3: Design of steel structures – part 1-13: Steel beams with large web openings. CEN/TC 250/SC 4 N 1839 - Web Openings in Steel Beams EC3 Part 1-13. European Committee for Standardization: Brussels, Belgium (2017)
8. Ward, J.K.: P100-Design of composite and non-composite cellular beams, ed. S.C. Institute (1990)
9. CEN: EN1993-1-2 Eurocode 3: Design of Steel Structures, Part 1-2: General rules, Structural fire design. European Committee for Standardization: Brussels, Belgium (2005)
10. Vila Real, P.M.M., et al.: Parametric analysis of the lateral–torsional buckling resistance of steel beams in case of fire. Fire Safety J. **42**(6), 416–424 (2007)
11. Demirdjian, S.: Stability of castellated beam webs. In Department of Civil Engineering and Applied Mechanics, McGill University, 140 p (1999)
12. ECCS: Manual on Stability of Steel Structures (nº 22) (1976)
13. ECCS: Ultimate Limit State Calculations of Sway Frames with Rigid Joints (nº 23) (1984)
14. Sonck, D., Boissonade, N., Van Impe, R.: Instabilities of cellular members loaded in bending or compression. In: Proceedings of the Annual Stability Conference Structural Research Council, p. 19 (2012)
15. Kwani, S., Wijaya, P.K.: Lateral torsional buckling of castellated beams analyzed using the collapse analysis. Procedia Eng. **171**, 813–820 (2017)

Wood Connections Under Fire Conditions Protected with Gypsum Plasterboard Types A and F

Elza M. M. Fonseca[1(✉)], Pedro A. S. Leite[2], and Lino Silva[2]

[1] LAETA, INEGI, Department of Mechanical Engineering, Instituto Politécnico do Porto, ISEP | Instituto Superior de Engenharia do Porto, Porto, Portugal
elz@isep.ipp.pt

[2] Department of Mechanical Engineering, Instituto Politécnico do Porto, ISEP | Instituto Superior de Engenharia do Porto, Porto, Portugal

Abstract. The wood behavior varies with temperature, which has a negative effect, being verified by progressive properties degradation. After fire exposure, it is possible to measure the wood char layer and the remaining resistant cross-section area. Based on numerical models and analytical equations, it is possible to study the protection level of wood connections, during fire exposure, and evaluate wood and steel dowels temperature fields in conjunction. The main objective of this work is to verify the fire resistance of unprotected (W-W-W) wood connections and compare the results with protected connections using gypsum plasterboard. According Eurocode 5, part 1-2 the connections could be protected by the addition of wood paneling, wood-based panels or gypsum plasterboard type A, H or F. To obtain conclusions according the application of different type of gypsum plasterboard in the designed W-W-W connections in different densities material, numerical models were building to verify the level of protection in fire action. Design methods require the use of analytical methodologies and computational modelling to predict the fire exposure and the components capacity to resist to this action. Results of the temperature field in the studied connections under fire will be presented, with the measured char layer in unprotected members and comparing the protected efficiency with different types of gypsum plasterboard.

Keywords: W-W-W connection · Fire · Protected connection · Charring rate

1 Introduction

The fire resistance of the wood connections depends on the char layer evolution, the strength reduction of the steel dowels and the residual cross section of the members. The fire performance analysis is complex because there are different materials involved and different parameters, geometries and dowels arrangements. Considering the behavior of wood members when submitted to a developing fire, they will burn and are rated as combustible [1, 2].

P. A. G. Piloto et al. (Eds.): CILASCI 2019, LNCE 1, pp. 93–106, 2020.
https://doi.org/10.1007/978-3-030-36240-9_7

The behavior of wood connections subjected to fire require more research. The assessment of the connection (wood member with steel fasteners) and their failure modes due its fire exposure is complex [3].

Wood material when exposed to fire presents a thermal physical degradation. The interface between the charred and noncharred wood is the transition phase between the black and brown material and is characterized by a threshold value of 300 [°C], according Eurocode 5 part 1-2 [4]. Also the thermal properties of wood vary considerably with temperature and should be defined according the Eurocode 5 part 1-2 [4]. This standard code provides the design values for the density, thermal conductivity and specific heat of wood.

According Eurocode 5, part 1-2 the wood connections could be protected by the addition of wood paneling, wood-based panels or gypsum plasterboard type A, H or F [4].

Raw materials of the gypsum plasterboards are gypsum, paper and additives. Gypsum consists of a porous core between two layers of paper, creating a sandwich structure. Usually, gypsum plasterboards have a thickness range between 9 to 30 [mm]. Gypsum is a non-combustible material and makes no contribution to fire, it works, in fact, as a built-in sprinkler [5].

Gypsum Type A corresponds to the conventional gypsum plasterboard with porous core and no reinforcement, except the paper laminated surface. Gypsum Type H with water-resistant properties. And Type F refers to 'fire-resistant' gypsum plasterboard with improved core cohesion at high temperatures.

This article first describes the basic idea of the design W-W-W connection designed at room temperature, based on Eurocode 1995, part 1.1, followed by connections fire exposure calculations, accordingly with Eurocode 1995, part 1-2. Birch was the studied wood specie, produced in glue laminated timber, grades GL20h, GL24h and GL32h. Studied dowels diameters were 6, 8, 10 and 12 [mm]. The connections were designed for three different applied loads, 10, 15 and 20 [kN].

Then, the main results of thermal and transient numerical analyses on unprotect connections and protected connections with gypsum plasterboards are presented. Numerical models, using a two-dimensional and transient thermal analysis, were developed in order to evaluate gypsum plasterboards cladding on W-W-W connections, which were designed with different wood densities. For unprotected connections under fire exposure, temperatures were evaluated, as well, the char layer and charring rate. That was followed by a comparison to Eurocode 1995 part 1-2 results in relation to different gypsum plasterboards studied, and at the end, steel dowels fields temperatures along the length were analyzed, after fire exposure.

The results will enlarge the knowledge of the fire behavior in unprotected and protected connections, that allow the calculation of the charring rate of the designed W-W-W models, the verification of the material density effect and the dowel diameter. This study carries out more results to the previous investigations developed by the authors [6–12].

2 Simplified Equations from Eurocode 5

2.1 Unprotected W-W-W Connections

The W-W-W connections were designed with simplified equations from Eurocode 5 part 1-1, at ambient temperature [7, 8, 10, 13]. Three types of wood material with different densities (GL20h, GL24h and GL32h) were used to building the W-W-W cnections with different steel dowels, as fastener members.

According Eurocode 5 part 1-1 [13], the design tensile strength along the grain $f_{t,0,d}$, must be equal or higher than the design tensile stress along the grain. The tensile strength represents a reduced value of the characteristic tensile strength along the wood grain, due to the application of safety factors: the modification factor for load duration and moisture content k_{mod}, and the partial factor for material properties γ_M, as the following expression:

$$f_{t,0,d} = \frac{k_{mod} \times f_{t,0,k}}{\gamma_M} \tag{1}$$

Considering E_d as the applied load and A_s the cross-section of the member, the design tensile stress along the grain $\sigma_{t,0,d}$, is according Eq. 2:

$$\sigma_{t,0,d} = \frac{E_d}{A_s} \tag{2}$$

According the simplified equations from Eurocode 5 part 1-1 [13], the characteristic load-carrying capacity, per shear plane and fastener, for a connection W-W-W with dowel fasteners, is determined according Eq. 3.

$$F_{v,Rk} = min\begin{cases} f_{h,1,k}\, t_1\, D \\ 0{,}5 f_{h,2,k}\, t_2\, D \\ 1{,}05\, \frac{f_{h,1,k}\, t_1\, D}{2+\beta}\left[\sqrt{2\beta(1+\beta) + \frac{4\beta(2+\beta)M_{y,Rk}}{f_{h,Rk}\, D\, t_1^2}} - \beta\right] + \frac{F_{\alpha x,Rk}}{4} \\ 1{,}15\sqrt{\frac{2\beta}{1+\beta}}\sqrt{2M_{y,Rk} f_{h,1,k}\, D} + \frac{F_{\alpha x,Rk}}{4} \end{cases} \tag{3}$$

t_1 represents the thickness of the wood members; $f_{h,1,k}$ is the characteristic embedment strength in timber member; D is the dowel diameter; $M_{y,Rk}$ is the characteristic yield moment of the fastener; $F_{\alpha x,Rk}$ represents the characteristic axial withdrawal capacity of the fastener and β is the ratio between the embedment strength of the members. The value of $M_{y,Rk}$ is calculated according the dowel diameter and the material strength of the bolt.

$$M_{y,Rk} = 0{,}3 f_{u,k}\, D^{2,6} \tag{4}$$

The value of the characteristic embedment strength in timber elements, is obtain by the value of the dowel diameter and the characteristic density of the wood ρ_k, as represented in Eq. 5.

$$f_{h,1,k} = 0{,}082(1 - 0{,}01D)\rho_k \tag{5}$$

With the calculation from $F_{v,Rk}$, it is possible to obtain the number of the bolts, Eq. 6.

$$N = \frac{E_d}{F_{v,Rd}} \tag{6}$$

The design equations are only valid if there is no premature splitting or shearing of the wood resulting in brittle type failure. To reduce the risk of such failures, minimum edge, end and spacing criteria for connections with dowel type fasteners have been developed and the procedure for bolts are given from the Eurocode 5 part 1-1 [13]. So, the designed connection at room temperature guarantees the applied load design.

With all design equations a worksheet was developed to allows the calculation of different W-W-W connections. Different parameters were considered: four dowel diameters, three applied tensile load and three wood materials. The obtained results allow to verify a linear correlation between the applied load and the necessary fasteners diameter, with a higher increase for lesser dowels diameter (see Fig. 1).

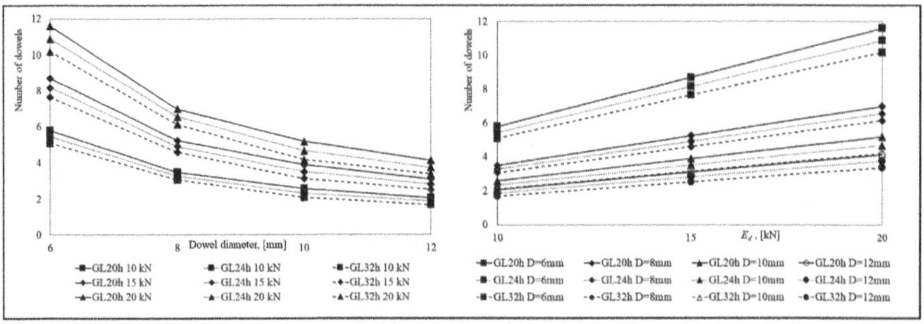

Fig. 1. Comparison between dowels diameter and number of dowels; applied load and number of dowels; for any type of three types wood material.

The required amount of dowels to use decreases significantly with increasing dowel diameter and increases with decreases the applied load. Wood density does not significantly change the mechanical strength of the connection, nevertheless lower wood density affects increasing lightly the number of dowels. All results of this work will complement previous research from the authors [6–12].

2.2 Protected W-W-W Connections

Simplified rules from Eurocode 5, part 1-2 [4] were used to analyze the same connections with side members of wood under standard fire exposure. The structural fire resistance in these type of wood material is equal to 0.7 [mm/min] in accordance to Eurocode 5, part 1-2 [4], that represents the design charring rate under standard fire exposure.

For connections with insulating material, the Eurocode 5 part 1-2 [4] proposes two options for the protection material: gypsum (type A, F or H) or wood-based panels. In this work, gypsum panel types A and F were chosen for protecting the wood connections. For gypsum type A:

$$t_{ch} \geq t_{req} - 0{,}5\, t_{d,fi} \tag{7}$$

For gypsum type F:

$$t_{ch} \geq t_{req} - 1{,}2\, t_{d,fi} \tag{8}$$

The value t_{ch} refers to the delay of start of charring rate due to protection. The time of the fire resistance $t_{d,fi}$, is according with the connector. For dowels, this time is 20 [min], however the minimum value for t_1 is 45 [mm]. t_{req} represents the required time of fire resistance. The fire protective panel thickness h_p is given according the Eq. 9, for both gypsum panel types:

$$h_p = \frac{t_{ch} + 14}{2{,}8} \tag{9}$$

In this work the protected connections, require a protection layer thickness of the panel equal to 18 [mm] for gypsum plasterboard type F and 23 [mm] for gypsum type A, assuming standard fire resistance period of 60 [min] and for any type of chosen wood densities.

3 Computational Model

3.1 Wood Connection

In addition, a numerical program was used to produce simulations focused on thermal and transient analysis to study W-W-W connections. This methodology allows to obtain the temperature field in the studied connections under fire, allows to calculate the char layer in the members with different wood densities, when unprotected, and compare the efficiency in protected connections with two different types of gypsum plasterboard.

Due to the symmetry of the geometry and the applied boundary conditions, the numerical calculation was performed for two dimensional (2D) plane of the connection using plane finite elements with 8 nodes. The size of the finite element is equal to 2 [mm] of length. The initial temperature in the numerical model was considered equal to 20 [°C]. The external surface of the connection is exposed to the standard fire curve ISO834 during one hour and the convection coefficient is taken equal to 25 [W/m^2 K] [14]. The surface emissivity is taken constant and equal to 1 for exposed side [14]. The non-linearity due to the thermal properties dependence was taken into account in the numerical simulation.

The numerical models were developed, using the designed connections obtained from the previous calculation at room temperature. Figure 2 presents a typical 3D W-W-W connection where the dashed green line represents the cross-section for 2D study through the dowels. According the symmetry of the connection, represented in blue line, only a quarter of the W-W-W connection will be studied. Figure 2 represents, in simultaneous, the 2D meshes used in the numerical model. In the represented meshes, the blue zone is the wood material, violet is the steel dowels and the red color the insulation material. For the protected connections (with additional color in red) a regular mesh was increased in depth, with h_p value for the fire protective panel thickness.

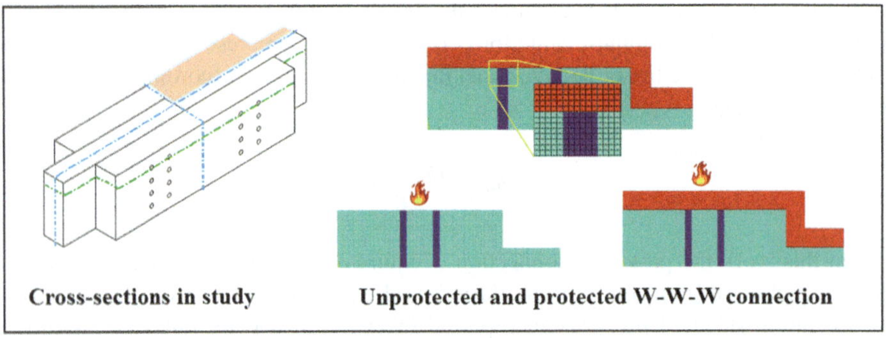

Fig. 2. W-W-W connection and 2D meshes for cross-sections.

3.2 Thermal Properties

Figures 3 presents all thermal properties (thermal conductivity, specific heat and density) used in the numerical model, based on standard codes and the literature: steel is according the Eurocode 3 part 1-2 [15], Eurocode 5 part 1-2 [4] is used for wood with different densities (GL20h, GL24h and GL32h) and each gypsum type is obtained from the literature, [16, 17]. The density for steel is constant, equal to 7850 [kg/m^3].

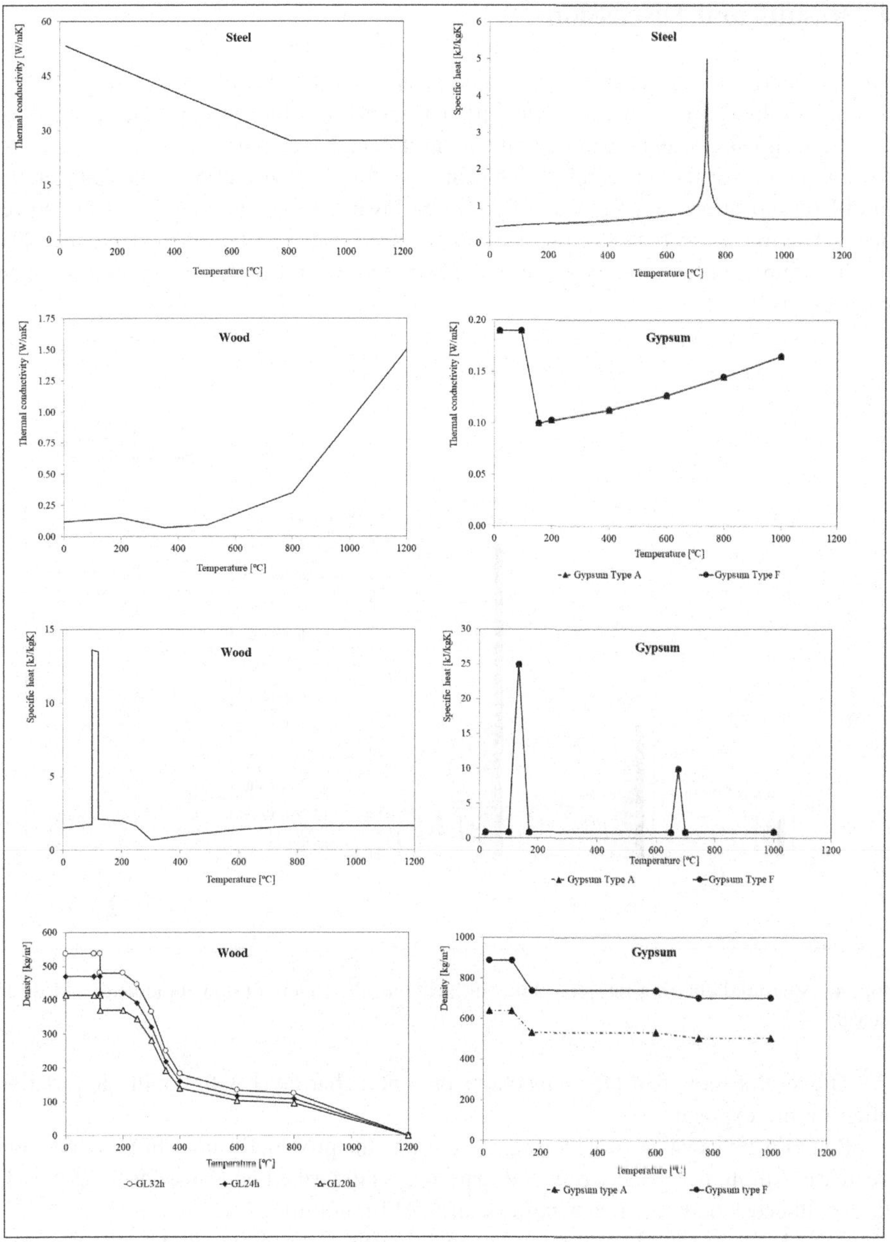

Fig. 3. Material properties.

4 Results and Discussion

The evaluation of the char layer thickness, depends on the fire exposed time, that determines the charring rate in [mm/min]. The char layer thickness was determined for each model, using different measurements in different locations.

Figure 4 presents the calculated charring rate for unprotected W-W-W connections in different locations: K1 faraway of the steel dowels, K2 in the vicinity of the dowel and K3 in the wood between two dowels, as represented in the located picture. The results comparing the three types of wood and the four dowel diameters used in wood connections.

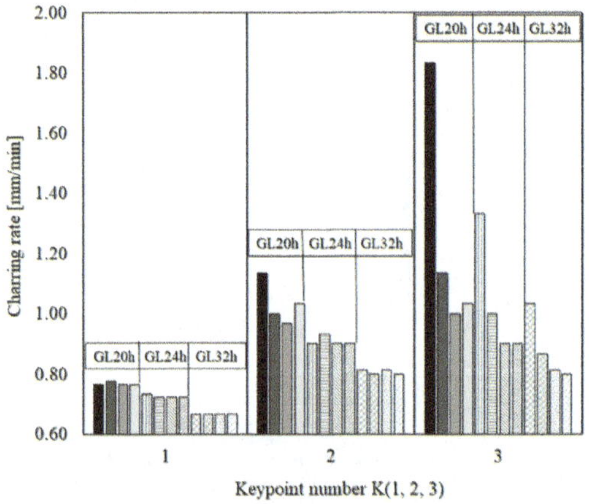

Charring rate [mm/min]

Wood, diameter	Keypoint number 1	2	3
GL20h d=6 mm	0.77	1.13	1.83
GL20h d=8 mm	0.78	1.00	1.13
GL20h d=10 mm	0.77	0.97	1.00
GL20h d=12 mm	0.77	1.03	1.03
GL24h d=6 mm	0.73	0.90	1.33
GL24h d=8 mm	0.72	0.93	1.00
GL24h d=10 mm	0.72	0.90	0.90
GL24h d=12 mm	0.72	0.90	0.90
GL32h d=6 mm	0.67	0.81	1.03
GL32h d=8 mm	0.67	0.80	0.87
GL32h d=10 mm	0.67	0.81	0.81
GL32h d=12 mm	0.67	0.80	0.80

Fig. 4. Measured charring rate in 3 different locations, dependent of dowel diameter and wood density.

The results show that the connections present a char layer with nonlinear variation after the fire exposure.

Point K1 gives values of charring rate close to the proposed values in Eurocode, but decrease with higher density of wood material. As described on Eurocode 1995 part 1-2, the studied wood has a charring rate of 0.70 [mm/min].

Point K2 is in the vicinity with the steel dowels and presents higher charring rate evolution. Point K3 presents also higher charring rate, close to the values from K2, but in wood connections in GL20h and lower dowel diameter of 6 [mm] presents the greater value.

The results showed that charring rate, measured between steels dowels, for 6 [mm] dowels diameter, was 1.83 [mm/min], 1.33 [mm/min] and 1.03 [mm/min], for GL20h, GL24h and GL32h, respectively.

The charring rate increased in the steel dowels vicinity, as shown for GL20h and 6 [mm] diameter connection, with the results, 0.77 [mm/min], 1.13 [mm/min] and 1.83 [mm/min], at measuring points, far from dowels, close to the dowels and between steel dowels, respectively.

Using only the Eurocode 5 part-1-2 [4] it is impossible to understand the fire effect through and inside the connection.

In general, there was an increase in charring rate related with the decreasing in wood density. The influence of the steel dowels on charring rate is also notable. The combination of the smallest wood density with the smallest bolt diameter is the worst condition for charring rate. In connections with lower dowels diameters, the number of fasteners need to be higher.

The previous analyze is very important because permits to verify the steel dowels diameter and wood density which provide a higher heat flux inside the connection. Figure 5 represents the time temperature history in unprotected and protected connections with gypsum plasterboard type A and F, with steel dowel 6 [mm] and GL20h, during one hour of fire exposure. The temperature results are presented in different points. K1, K2, K3 and K4 are presented in front of fire in unprotected connections and in the interface of the gypsum plasterboard for protected connections. More four points (K5, K6, K7 and K8) represent the temperature inside the unprotected connection.

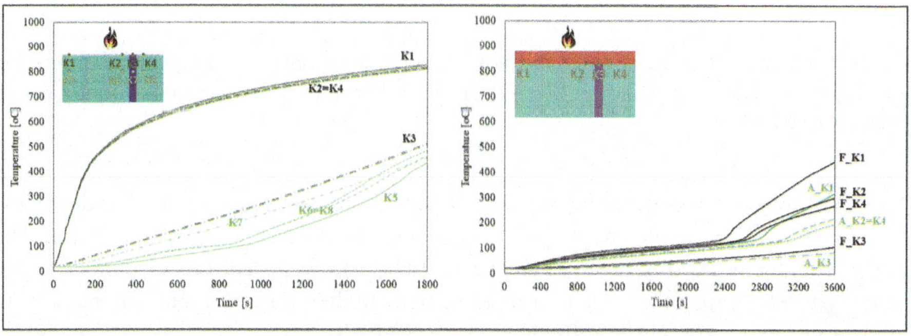

Fig. 5. Time-temperature history in different locations for unprotected and protected connection GL20h with dowel diameter equal to 6 [mm].

For unprotected connections, the temperature in K1, K2 and K4, follows the standard fire curve, and K3 presents lesser temperature during the fire time exposure. The inside measured points K5, K6, K7 and K8 have lower temperature through the time. The measuring in points K7 and K3 are quite similar that demonstrate the steel conductivity which remains equal. Steel, an excellent heat conductor, and by the action of fire, influences the heat conduction to the wood with intensity. The area of the wood surrounding the steel dowel thus increases in temperature and may even carbonize when temperature reaches 300 [°C] after 1000 [s]. In simultaneous, the wood allows a decrease in temperature along the steel dowels, making the effect of fire not so pronounced, thus having an insulating role.

For protected connections gypsum type F causes more heat through the connection. Nevertheless, the temperatures through one hour of fire exposure are always lesser than 300 [°C], excepting the point F_K1. The gypsum plasterboard proved to be a good thermal insulator. The gypsum material allows to vary the thermal insulant effect, reducing the wood char layer in the connections, due their wide range of physical properties.

Figure 6 represents the temperature comparison along the length of the steel dowels for unprotected and protected connections with gypsum plasterboard type A and F for dowel diameter 6 [mm] and GL20h, the worst situation according the previous conclusions.

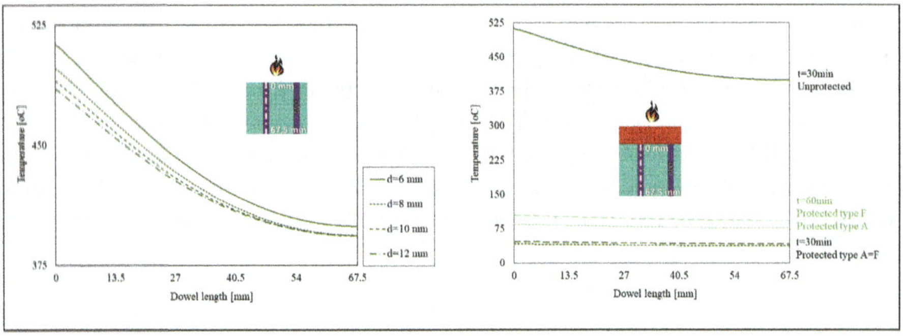

Fig. 6. Measured temperature along the dowel, unprotected connection (with all diameters and time exposure of 30 [min]) and protected (diameter 6 [mm], time exposure of 30 [min] and 60 [min]), wood GL20h.

In unprotected wood connection, the temperature decreases to the inside of the connection, at time exposure fire 30 [min], between side to side of the dowel. The medium temperature difference could be 100 [°C]. Lower steel dowel diameter promotes greater temperature when compared with higher dowels diameter, namely in front of fire exposure with a difference in 25–30 [°C].

In protected connections, the temperature is constant along the dowels and is low due to the insulation of the plasterboard and the low thermal conductivity of the wood, for any time of fire exposure (30 or 60 [min]). The protection produces lower temperatures inside the connection with less values than 100 [°C]. Gypsum type F promotes more heat inside the connection, in general around of 25 °C than type A.

Figure 7 represents the comparison between char layer on the unprotected wood during 30 [min] of fire exposure and the protect effect using gypsum panels during 60 [min] of fire. The results were presented only for wood density GL20h in connections with dowel diameter equal to 6 [mm], which promote more heat inside the connection. The char layer on the wood is represent in grey color. Grey color also represents the thickness of the gypsum panel (A or F) when reaches 300 [°C].

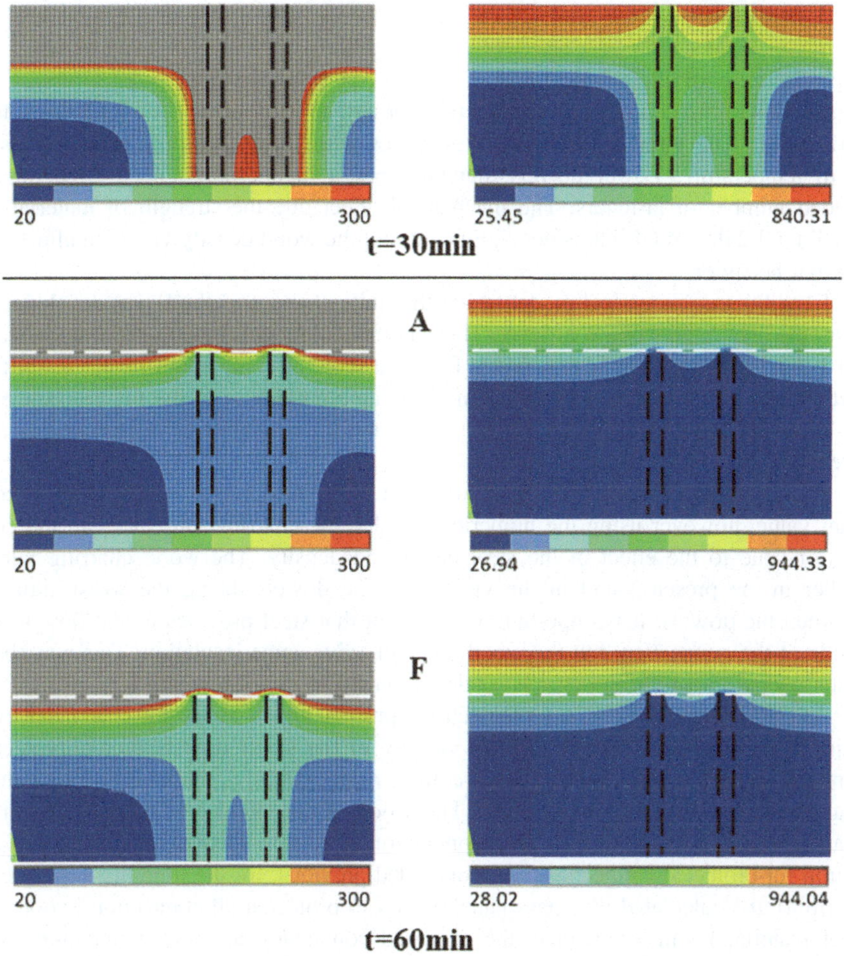

Fig. 7. Temperature in unprotected connection (exposure time of 30 [min]) and protected connection (exposure time of 60 [min]), wood GL20h and dowel diameter 6 [mm].

According the results, the unprotected connection within 30 [min] is severely broken around the connectors. The gypsum plasterboard maintains the integrity of the connection for a period of 60 [min]. Every protected connection is prepared analytically to resist during 60 [min] at fire exposure. Gypsum type F has a lesser thickness, and in generally protect the wood, but produces higher temperatures. In general, both gypsum panels are a good thermal protector of the wood connections, during 60 [min] of fire.

In summary, the wood charring rate decreased with its rising density and it increased with steel dowels vicinity. Gypsum plasterboards showed to be a good solution to protect W-W-W connection exposed to fire, and type F gypsum was associated to a lower thickness compared to type A, for the same protection time.

5 Conclusions

A procedure with all analytical and simplified equations were presented to assess the dimensions of the connection cross-section for any applied tensile load, and fire safety verification. The number of fasteners increases with load, according to standards at room temperature. Lower dowels diameter, has a higher pronounced effect in the needed number of fasteners. The mechanical effect due the strength of material for GL20h, GL24h and GL32h is not significant, but the wood density variation affects the thermal behavior.

In addition, a numerical model, using the finite element method, was developed in order the obtain the wood char layer. These numerical models help to better understand the behavior of the connection under fire exposure, in comparison to the simplified method. The numerical model gives an acceptable prediction of the fire resistance of W-W-W connections, and allows to determine the fire resistance of different types and sizes of connections.

According the Eurocode 5, the charring rate is considered as a standard and constant value, however using the numerical results, the charring rate varies in the connection, due to the effect of the steel and wood density. The wood charring rate is higher in the presence and in the vicinity to the dowels, being the worst situation between the dowels. It is important to point out that steel provides a heat flow to the inside of the connection, but the wood members give some insulation. This way, both materials participate in the char layer development in the connection.

After the fire exposure, the connections present higher values of char layer evolution with nonlinear variation in the vicinity of the steels dowels. For unprotected connections with a time of fire exposure greater than 1000 [s], the steel dowels induce heating inside the wood core section. The wood in the vicinity of dowels presents a char layer that is unfeasible for the connection. Due this situation will be better the option for protected connections. As concluded previous, for the time fire exposure of 60 [min], the calculated thickness panel allows to protected all connection (wood and steel materials) remaining intact the designed connection at room temperature. The gypsum material is a good thermal insulant reducing the char layer compared to the unprotected connections.

The numerical method, both for the protected and unprotected connections, shows a good process to understand the behavior of the connection under fire exposure, an enhanced methodology when compared with the analytical results obtained from the simplified equations. As a general conclusion, the combination of the largest wood density with the largest dowel diameter allows lower temperatures for the connection in a fire situation. In opposite, a lower wood density and a smaller dowel diameter lead to a higher charring rate due the greater number of connectors.

Acknowledgments. The basis for this article is a paper which we prepared for the 5CILASCI - *5° Congresso Ibero-Latino-Americano em Segurança Contra Incêndios* [8], updated to reflect new insights and actions since the conference. We thank the 5CILASCI for allowing us to republish some of our research in this journal.

Funding. This research received no specific grant from any funding agency in the public, commercial, or not-for-profit sectors.

Conflict of interest statement. The author(s) declare no conflicts of interest associate with the presented work.

References

1. Fonseca, E.M.M., Coelho, D.C.S., Barreira, L.M.S.: Structural safety in wooden beams under thermal and mechanical loading conditions. Int. J. Saf. Secur. Eng. **2**(3), 242–255 (2012)
2. Fonseca, E.M.M., Barreira, L.M.S.: Experimental and numerical method for determining wood char-layer at high temperatures due an anaerobic heating. Int. J. Saf. Secur. Eng. **1**(1), 65–76 (2011)
3. Peng, L., Hadjisophocleous, G., Mehaffey, J., Mohammad, M.: On the fire performance of double-shear timber connections. Fire Saf. Sci. **42**, 1207–1218 (2011)
4. CEN, EN1995-1-2: Eurocode 5: Design of timber structures. Part 1-2: General Structural fire design, Brussels (2004)
5. Just, A., Schmid, J., König, J.: Structural Gypsum plasterboards used as fire protection - analysis of a database. SP Technical Research Institute of Sweden, 29, SP Report (2010)
6. Martins, R., Fonseca, E.M.M.: Fire behaviour of protected W-S-W connections with a steel plate as the central member and different dowels diameter. Int. J. Sci. Technol. **4**(3), 60–78 (2018)
7. Fonseca, E.M.M., Silva, L., Leite, P.A.: Fire safety of wood-steel connections. In: Amélia, M., Barbosa, J., et al. (eds.) 4th International Conference on Numerical and Symbolic Computation Developments and Applications – Developments and Applications, pp. 109-118. Edited by APMTAC, Porto (2019)
8. Leite, P.A.S., Silva, L., Fonseca, E.M.M.: The fire resistance of (W-W-W) wood-to-wood connections protected with different types of gypsum plasterboard. In: Piloto, Ferreira, Fonseca, Baptista, Castro, Mesquita, Vaz, Chichorro, Guedes, et al. (eds.) 5° Congresso Ibero-Latino-Americano em Segurança Contra Incêndios, pp. 543-552. Edited by ALBRASCI., Porto (2019)
9. Silva, L., Leite, P.A.S., Fonseca, E.M.M.: The Density Effect in (W-S-W) wood connections with internal steel plate and passive protection under fire. In: Piloto, Ferreira, Fonseca, Baptista, Castro, Mesquita, Vaz, Chichorro, Guedes, et al. (eds.) 5° Congresso Ibero-Latino-Americano em Segurança Contra Incêndios, pp. 563–571. Edited by ALBRASCI, Porto (2019)
10. Aissa, A., Fonseca, E.M.M., Daniel, A.P.M.: W-W-W connections in double-shear at ambient temperature: effect of the applied tensile load and dowels diameter. In: Gomes, M. (ed.) 6th International Conference on Integrity Reliability Failure, pp. 1349 1356. Edited by: INEGI/FEUP, Lisboa (2018)
11. Martins, D.A.R., Fonseca, E.M.M.: Double-shear W-S-W connections at ambient temperature, with different applied tensile loads and steel dowels diameter. In: Gomes, Meguid (eds.) 6th International Conference on Integrity-Reliability-Failure, pp. 1341–1348. Edited by: INEGI/FEUP, Lisboa (2018)
12. Martins, R., Fonseca, E.M.M.: W-S-W Connections with a steel plate as the central member and different dowels diameter at high temperature. In: Gomes (ed.) 1st Iberic Conference on Theoretical and Experimental Mechanics and Materials/11th National Congress on Experimental Mechanics, pp. 239–248. Edited by: INEGI/FEUP, Porto (2018)

13. CEN, EN1995-1-1: Eurocode 5: Design of timber structures. Part 1-1: General Common rules and rules for buildings, Brussels (2004)
14. CEN, EN1991-1-2: Eurocode 1: Action on structures. Part 1-2: General actions - Actions on Structures Exposed to Fire, Brussels (2002)
15. CEN, EN1993-1-2: Eurocode 3: Design of steel structures. Part 1-2: General rules - Structural Fire Design, Brussels (2005)
16. Rahmanian, I., Wang, Y.: Thermal conductivity of gypsum at high temperatures a combined experimental and numerical approach. Acta Polytechnica **49**(1), 16–20 (2009)
17. Frangi, A., Schleifer, V., Fontana, M., Hugi, E.: Experimental and numerical analysis of gypsum plasterboards in fire. Fire Technol. **46**, 149–167 (2010)

Emergency Exits: Analysis and Reflection Based on a Modelling and Standardization Study

Edna Moura Pinto(✉) and Mariana Lima Oliveira Montenegro

Universidade Federal do Rio Grande do Norte, Natal, Brazil
emourap@ufrnet.br

Abstract. Safe evacuation of a building is favored by complying with regulatory guidelines, it is also closely related to accessibility that must be fully resolved in the architectural project, being directly associated with the space distribution and the horizontal and vertical circulation. The relation between accessibility and the spaces in a building can be explained by the Social Logic of Space Theory or Space Syntax Theory. The aim of this study is to identify and analyse possible divergences in the prescribed standards for fire evacuation adopted in the State of Rio Grande do Norte-Brazil, based on the evaluation of design solutions oriented by standards in conjunction with computer simulations. The assessments were carried out using the floor plans and two softwares: PATHFINDER, used to relate the escape time to the moving of people during the egress; and DEPTHMAP used to relate the design to the space configuration study, its relation to the user's movement, and the placement of emergency exits in the building. It was concluded that the NBR 9077:2001 and the TI 11:2014, while efficient for the most part, present lacks, with respect to the dimensioning of stairs, number of maximum occupants, and maximum distance travelled to a safe location.

Keywords: Architecture · Fire safety · Emergency exits · Space syntax

1 Introduction

Fire Safety (FS) in Brazil is conducted by prescribed standards issued by the Brazilian Association of Technical Standards (Associação Brasileira de Normas Técnicas – heretofore, ABNT). Each State in Brazil regulates the mandatory requirements on which the actions of the auditing agencies are based. This fact contributes to the non-uniform protocols on a national level, that aligned with the constant need to update the standards, has motivated us to analyze some projects that offer solutions that have been adopted in Rio Grande do Norte, Brazil, based on standardization.

Safe evacuation of a building is favoured by complying with regulatory guidelines in effect, however, it is also closely related to accessibility, which, according to Brentano [1] is characterized as: "The possibility and reach condition, perception and understanding for the safe and autonomous use of buildings, spaces […] etc.".

P. A. G. Piloto et al. (Eds.): CILASCI 2019, LNCE 1, pp. 107–120, 2020.
https://doi.org/10.1007/978-3-030-36240-9_8

Accessibility must be fully resolved in the architectural project, being directly associated with the space distribution and the horizontal and vertical circulation, which are not clearly presented in the standards.

The relation between accessibility and the spaces in a building can be explained by the Social Logic of Space Theory or Space Syntax Theory [2]. "Space here is understood as a barrier system and permeabilities, i.e. accessible and inaccessible areas to our movement." [3] Depending on how the space is configured, it facilitates, hinders or even blocks routes. Ideally, the space configuration should facilitate moving and directing people during an evacuation, thus the close relationship between Space Syntax and Fire Safety.

Currently, computational methods provide assessments about solutions implemented in the architectural designs and allow us to reflect on the standards adopted, and to indicate improvements and updates. The analyses carried out as part of a master thesis [4], address design solutions for emergency exits at public assembly buildings. The results allow us to reflect on projects that are informed by the standard guidelines for these types of buildings.

2 Objective

The objective of this study is to identify and analyse possible divergences in the prescribed standards for fire evacuation adopted in the State of Rio Grande do Norte-Brazil, based on the evaluation of design solutions oriented by standards in conjunction with computer simulations.

3 Methods

The analysis of the floor plan of three academic facility buildings, characterized as public assembly buildings, were carried out using AUTOCAD® [5] software, based on fire safety standards in Rio Grande do Norte – RN: NBR 9077 [6], the Panic and Fire Prevention and Safety Code of the Rio Grande do Norte – RN (heretofore, as in the Portuguese, CBMRN) [7] and the Technical Instruction 11 (TI 11)- Emergency exit [8].

The assessments were carried out using the floor plans and two softwares: PATHFINDER® [9], used to relate the escape time to the moving of people during the egress; and DEPTHMAP® [10], used to relate the design to the space configuration study, its relation to the user's movement, and the placement of emergency exits in the building. Through the Pathfinder—a simulator software developed by Thunderhead Engineering Consultants – it was possible to: simulate the evacuation and movement of people in the buildings; predict the escape time; and to point out critical areas in this process, such as the agglomeration spots. The second software is related to Space Syntax, which allowed for the analysis of the spatial configuration, and a description of how it is related to the tendencies of the occupants to move through space. The Depthmap was developed at University College London to perform a set of spatial network analyses designed to understand social processes within the built environment, the aim of the software is to produce a map of open space elements, connect them via some relationship and then perform graph analysis of the resulting network. The objective of the analysis is to derive variables which may have social or experiential significance. It was verified whether the space configurations were intelligible, and if the emergency exits were accessible to users in the entire building, as well as identifying the accessibility so that the occupant could reach an emergency exit or a safe location, analysing thus the placement of these exits.

The data collected from the simulations in both of the softwares were compared to each other and with the previously cited prescribed standards.

4 Simulations and Results

Four scenarios were modeled for the simulations: (1) Conventional stairs – without protection; (2) One inactive stairway (Buildings 1 and 2) or with width modifications (Building 3); (3) Closed stairs – protected; (4) Consideration of an extra emergency exit on the last floor or to the rooftop. The simulation of scenario 2 using Depthmap was not carried out on Building 3, given that this scenario did not accrue changes in the results of the Statistical Analysis of the Space.

In the first case, Building 1, according to the CBMRN, the building fulfilled all of the requirements listed; however, in relation to NBR 9077 there were some disparities: on the width of the stairs, in the hallways of the ground and first floor; on the maximum distance to be travelled to the externa area of the building on the first second and third floors; on the type of stairs; and on the presence of a refuge space. Results presented in Table 1.

In the analysis of Building 2, the parameters established by the TI 11 were attended to; however, the width of the stairs, the maximum distance to be travelled to the external area of the building on the first, second and third floor, the type of stairs and the presence of a refuge space according to the NBR 9077 were considered faulty or non-existent.

Finally, Building 3, according to the CBMRN, did not properly address the issue of the width of the stairs; and in relation to NBR 9077 there were also some irregularities: the width of the stairs and the exit doors to the external area, the maximum distance to be travelled to the external area of the building on the first and second floor and the type of stairs.

Subsequently, after the data collection concerning (non)conformity to the prescribed standards in the three buildings, the computational analyses were carried out.

In the Pathfinder simulations, researchers adopted the parameters described in Table 2. The walking speed was 1.3 m/s.

Table 1. Data on the Building from the standards analysis 1 [4].

Standard	Parameter		In project	Status
–	Total area of construction		6.841,40 m^2	–
NBR 9077	Building classification	Occupation	Scholastic (E-1)	–
		Height	M-Building of medium height (6,00 < H ≤ 12,00)	–
		Dimensions on blue-print	Q – all floors R – with a small underground W – Very large building (A > 5.000 m^2)	–
		Constructive characteristics	Y – Building with medium fire resistance	–
	Calculation of the population – Table 5 of the norm or layout	Underground	13 people	–
		Ground level	1.091 people	–
		1st floor	809 people	–
		2nd floor	392 people	–
		3rd floor	146 people	–
		Total	2.438 people	–
	Stair width–Unit passage capacity (u. p.)[a] = 60 (809 ÷ 60 = 14 u.p.)/Minimum width = 7,70 m (total value to be distributed between stairs)		All 3 stairs with 2,20 m width – total 6,60 m = 12 u.p.	F
	Width of outlets (access/hallways)[b] Unit passage capacity (u.p.) = 100 Ground floor = 6,05 m/1st floor = 4,95 m/2nd floor = 2,20 m/3rd floor = 1,10 m		Ground floor = 3,30 m 1st floor = 2,85 m 2nd floor = 2,85 m 3rd floor = 2,70 m	F F A A
	Width of ground floor doors – UP capacity = 100 (1.091 ÷ 100 = 11 u. p.)/minimum width = 6,05 m (total value to be distributed between the doors of the emergency exit)		6 doors with 3,85 m of width and 1 door with 2,00 m = total of 25,10 m = 46 u.p.	A
	Maximum horizontal distance to be travelled to a safe location (external area of the building) - Maximum = 45 m		Ground floor = 32,0 m 1st floor = 71,60 m 2nd floor = 85,10 m 3rd floor = 94,60 m	A F F F
	Number of exits on the outlet floor – minimum 2		7 exits	A
	Number and types of stairways – 2 PS (protected stairs)		3 stairs NP (not protected)	A NA
	Refuge area – Mandatory in educational buildings – AND – when classified as "W"		Not present	NA

(*continued*)

Table 1. *(continued)*

Standard	Parameter	In project	Status
CBMRN/IT 11	Width of the stairs– Unit passage capacity (u.p.) = 75 (809 ÷ 75 = 11 u.p.)/Minimum width = 6,05 m (total value to be distributed between stairs)	All 3 stairs with 2,20 m of width – total of 6,60 m = 12 u.p.	A
	Maximum distance to be travelled to the stairs – from the door of the unit farthest from the internal walkway < 10 m Ground floor = 75 m Other floors = 65 m	Ground = 32,00 m 1st floor = 48,00 m 2nd floor = 46,00 m 3rd floor = 46,00 m	A A A A
	Types of Stairs – NP	NP (not protected)	A

[a]Unit of passage is the minimum width for a row of people to pass (this value is set at 0.55 m)
[b]Author's note

Table 2. Input parameters to Pathfinder simulations.

Simulation mode and movement	Scenarios	General parameter
Steering	1, 2, 3 and 4	– All occupants go to nearest exit – Furniture according to the design layout – Number of occupants distributed throughout the environments according to the existing layout of the floor plan or calculated based on the standards – Population randomly included in the environments – All exits open – Standard pre-movement time established by the Pathfinder – All occupants walk at the same speed - 1.3 m/s

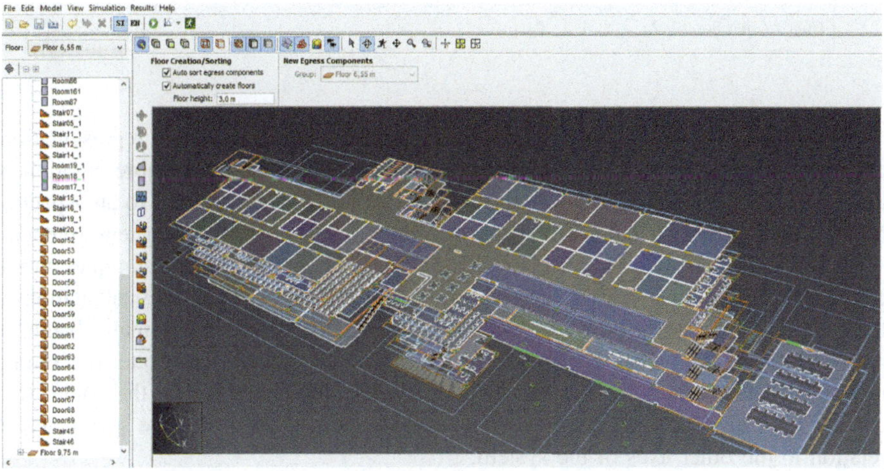

Fig. 1. Building 1 – floor plan during the Pathfinder simulation [4].

Fig. 2. Movement of the occupants using Pathfinder software.

At the end of each simulation (see Figs. 1 and 2), the Pathfinder generates a result report table. Table 3 presents the values generated for Building 3 in scenario 1, for example.

Table 3. Result report of a Pathfinder simulation [4].

Simulation:	study 3 scenario 1			
Mode:	Steering			
Total Occupants:	1.210			
Exit Times (s):				
Min:	1,5			
Max:	**231,8**			
[Components] All:	259			
[Components] Doors:	132			
Startup Time:	0,1s			
ROOM / DOOR	**FIRST IN** (s)	**LAST OUT** (s)	TOTAL USE (pers)	FLOW AVG. (pers)
Floor 0,40 m-> Door04	3,2	**206,8**	626	3,74
Floor 0,40 m-> Stair11	4,5	**188,3**	180	0,85
Floor 7,4 m ->Door101	3,8	5,8	3	1,52
Floor 7,4 m-> Room107	0,0	5,8	3	

Next, in the simulations using Depthmap, two of the techniques of representation, quantification and analysis of the Spatial Syntax were used: a linear representation of the space and a VGA (Visibility graph analysis). Each one was related to an aspect of how people experience and uses the space: people move in lines – linear representation – and dominate a visual field from any determined point – VGA. The techniques aim to characterize different levels of accessibility in the building through the calculation of the relation to access, whose numerical values are converted to chromatic bands in representations designated as axial maps and maps of visibility. The hotter (becoming red) indicate axes that are better connected or more integrated; while the colder (becoming dark blue) indicate axes less connected, less integrated or more segregated in relation to the other axes of the system.

The topological distance, which considers as the shortest path is that which permits the least number of changes in direction between the pair of points analyzed, related with the intelligibility of a spatial structure, once the "topological focuses properties such as the position, form of connection, relationship to the neighbors, adjacency, among others; and not the dimensional or metric properties" [12]. The theory defends that the space generates measurable properties, from the representation of the floor plans in maps or graphs. The topological measurements calculated in the present study were: Connectivity, Integration, Intelligibility, Depth and Visibility. See Figs. 3, 4, 5, 6

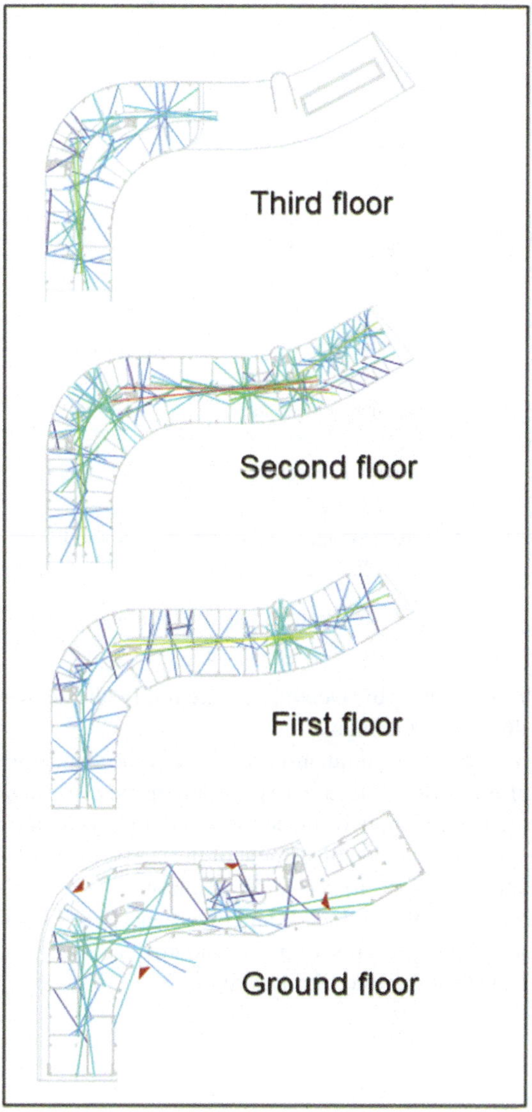

Fig. 3. Connectivity map for Building 2.

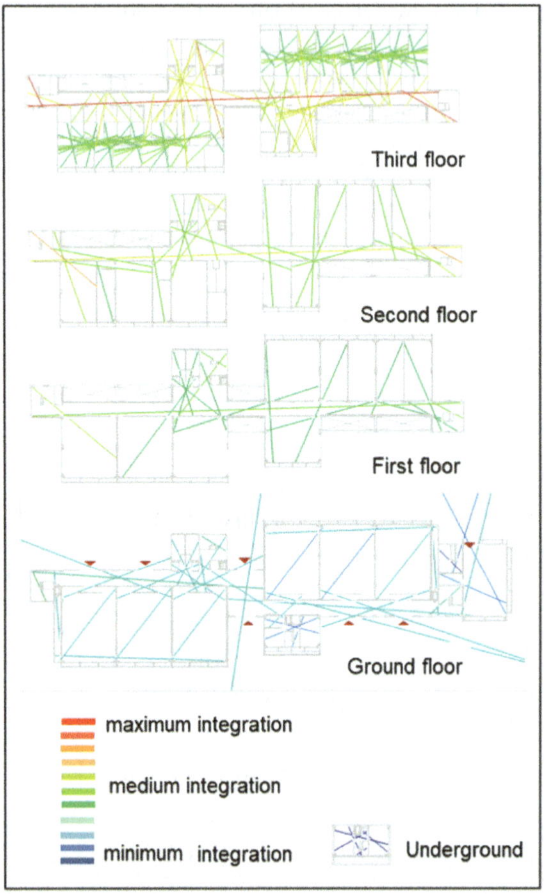

Fig. 4. Integration map for Building 1.

and 7, with results generated by Depthmap, exemplify the representations of these measurements in the three buildings.

The interaction between the simulation results and the standards was carried out to achieve the aims of this work, and to verify the emergency exit parameters previously cited in this research: maximum distance travelled to a safe location; number and dimensions of the emergency exits; placement of the stairs; total egress time for users in a building; occupant movement aspects – see Table 4.

Thus, the presence of both confluence and divergence was found within the results, given the proposal of the precepts of the standards, as the research presupposed. Table 5 contains a synthesis of the ultimate findings.

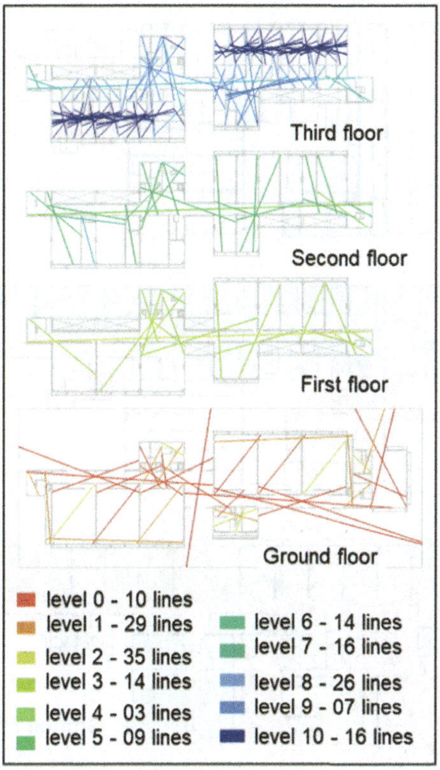

Fig. 5. Depth map for Building 1.

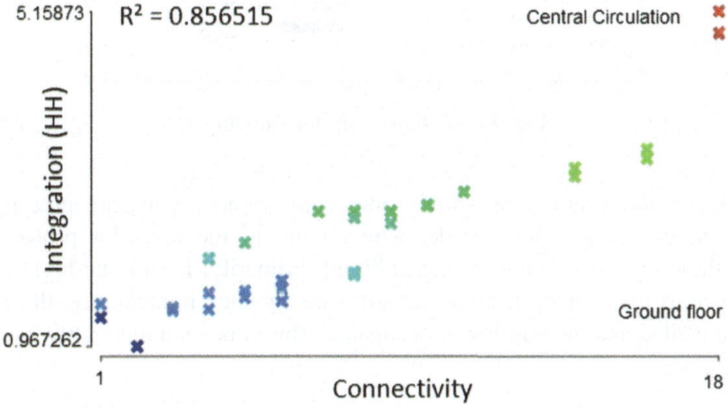

Fig. 6. Intelligibility - correlation graphic between integration versus connectivity.

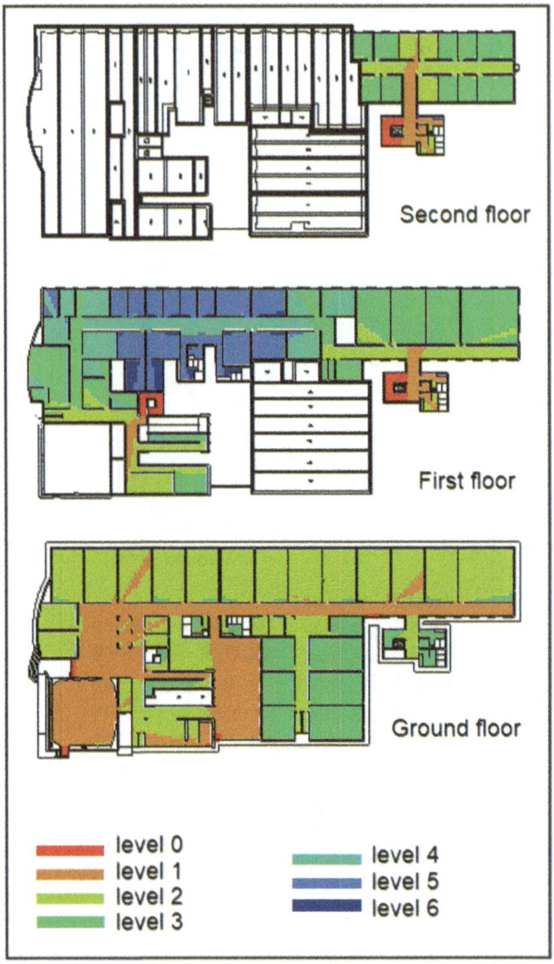

Fig. 7. Visibility map for Building 3.

Results of the simulations show some convergences with the precepts of the standards in design decisions made, which point to the need for protected stairs, however these are not verified in the building. Primarily, it must be highlighted that diverging from the standardization, the software signaled unfavorably, the maximum distance travelled and the number of occupants. This concurs unfavorably on the safety of the building.

Table 4. Interaction between the emergency exits and simulations results.

Software	Divergences	Convergences
PATHFINDER	**1**. In the quantity of occupants allowed by the standard (base on the area) for the dimensioning of the emergency exits (as it caused the crowding and increased the total escape time) **2**. In the maximum distance travelled to a safe location (protected stairway), given that the distance travelled on the ground floor to reach the external area of the building must also be added. **3**. In the need for a refuge area (as stipulated in NBR 9077:2001)	Need for protected stairways (NBR 9077) to improve the total escape time to a "safe location"
DEPTHMAP		The placement of the emergency exits (doors and stairs) were shown to be accessible in the spatial configuration of the buildings, even without having been detailed by the prescribed standards.

1 – The term "safe location" is in quotes due to the commentaries previously provided on prior topic with respect to this nomenclature in the NBR 9077:2001

Table 5. Synthesis of the results.

Parameter	What the Standards stipulate on these issues	Results/Conclusions	Qualification	Software
Maximum distance to be travelled to a safe place	NBR 90 77: 45 m to safe location and IT 11: 65 m to stairs	Studies 1, 2 and 3: • Buildings must follow mandates of NBR 9077 in relation to the protected stairs to conform to the maximum distances suggested; or revisit the term "safe location", thereby, reducing the total escape time. • The standards must also take into consideration the distance a person travels through the ground floor after leaving the protected stairway. It is not enough to count a safe distance to one of the protected stairways.	Unfavorable	PATHFINDER

(*continued*)

Table 5. (*continued*)

Parameter	What the Standards stipulate on these issues	Results/Conclusions	Qualification	Software
Quantity and dimension of the emergency exits	Three stairways and seven exits to the external area with dimensions conforming to the standards (in terms of height, of the dimensions on the blue-print and the constructive building characteristic)	Studies 1, 2 e 3: The number of occupants is taken into consideration just through the areas of the environments (dimensions in blue-print); with this, the quantity of people on each floor does not assure favorable results, as it was verified that there was too much crowding in the circulation areas and stairs.	Unfavorable	PATHFINDER
		Studies 1, 2 and 3: Positive results in relation to the topological distances and spatial configuration of the building	Favorable	DEPTHMAP
Position of the stairs	Reference just in terms of maximum distances to be travelled or distances between two stairways.	Studies 1, 2 and 3: • Despite the total changes of Direction necessary to reach a safe location having been high in the majority of the scenarios, many environments are considered with up to three changes. This can be seen as a positive result given the familiarity that the majority of people involved have in relation to the buildings. • Findings that show the importance of the axes with greater potential for movement are connected to the escape locations (stairs or door to the outside)) • Spatial configurations favourable to accessibility • Even without making direct reference to the placement of the stairs, the Standards are shown to be efficient.	Favorable	DEPTHMAP
		Study 2: The axes traced show the possibility of a person in transit moving to blocked hallways.	Unfavorable	
Total egress time for a user of the building	No mention in this respect	Compared with the time in NBS (2,5 min), the more favourable situation was with the protected stairs indicated by the NBR 9077	Favorable	PATHFINDER
Aspects of movement of the occupant	No mention in this respect	The tendency to move through the axes with greater integration was verified, throughout the axes that pass by environments with a greater number of occupants and by more connected axes. It was verified that there was a tendency to crowding in this case.	Unfavorable	DEPTHMAP and PATHFINDER

5 Conclusions

Despite the challenges of synthesizing many results generated by two distinct softwares with the standards, the research is seen to have achieved its objective. The analyses carried out contribute to verifying the performance of the three public assembly buildings with regard to the fulfillment or not, of the prescribed standards for emergency exits.

The usefulness of the Pathfinder and Depthmap softwares, is not limited to the procedures carried out in this research. This study is the beginning of a sequence of studies that can unfold, intending to continue to verify the validation of the proposal for uniformly applying prescribed standards; evaluate escape time, steering in case of fire, or the placement of emergency exits.

In a brief comparison between NBR 9077:2001 and TI 11:2014, some differences were found in their application. Although NBR 9077 is adopted all over Brazil and is applied more often, TI is updated, corrected and adapted more frequently. This leads us to hold that these latter characteristics are favorable and make it a more secure resource to be applied.

It is concluded that the NBR 9077:2001 and the TI 11:2014, while efficient for the most part, present lacks, with respect to the dimensioning of stairs, number of maximum occupants, and maximum distance travelled to a safe location, that can jeopardize the safe evacuation of the building. However, there is a greater number of data that can prove these affirmations and point to the problems with greater assurance, it is suggested that other simulations, on a higher number of buildings, must be considered and carried out in future researches, so that a greater sample and base for the study of this subject is increasingly more reliable and solid.

References

1. Brentano, T.: A Proteção Contra Incêndio no Projeto de Edificações, 2 edn. Telmo Brentano, Porto Alegre (2010)
2. Hillier, B., Hanson, J.: The Social Logic of Space. Cambridge University Press, Cambridge (1984)
3. Holanda, F. (org.): Arquitetura & Urbanidade. ProEditores Associados Ltda., São Paulo (2003)
4. Montenegro, M.L.O.: Análise de desempenho das saídas de emergência por meio de simulações computacionais - O caso de projetos de edifícios universitários. Dissertação (Mestrado em Arquitetura e Urbanismo) – Programa de Pós-Graduação em Arquitetura e Urbanismo, Universidade Federal do Rio Grande do Norte, Natal, 177p. (2016)
5. Autodesk Inc.: AUTOCAD, San Rafael, California, USA (2013)
6. Associação Brasileira De Normas Técnicas. NBR 9077: Saídas de emergência em edificações. ABNT, Rio de Janeiro (2001)
7. Governo Do Estado Do Rio Grande Do Norte: Corpo de Bombeiros Militar do Rio Grande do Norte. Código de Segurança e Prevenção contra Incêndio e Pânico do Estado do Rio Grande do Norte. [Imprensa Oficial], Natal (1979)

8. Secretaria De Estado Dos Negócios Da Segurança Pública: Corpo de Bombeiros da Polícia Militar do Estado de São Paulo. Instrução técnica N° 11 – "Saídas de emergência". São Paulo. [s.n.] (2011)
9. Thunderhead Engineering Consultants, Inc.: "Pathfinder", Manhattan, USA (2014)
10. University College London: Space Syntax Laboratory. "Depthmap", London (2014)
11. Ono, R.: O impacto do método de dimensionamento das saídas de emergência sobre o projeto arquitetônico de edifícios altos: uma análise crítica e proposta de aprimoramento. 2010. Tese (Livre Docência) – Curso de Arquitetura e Urbanismo, Universidade de São Paulo, São Paulo (2010)
12. Figueiredo, L.: Linhas de Continuidade no Sistema Axial. Universidade Federal de Pernambuco. Mimeo, Recife (2004)

Performance of Plaster Walls Exposed to High Temperatures

Roberta Tabaczenski de Sá[1(✉)], Cristovão J. D. Feitósa[1], José J. Bezerra[2], Tiago A. C. Pires[1], José J. Rêgo Silva[1], and Cleandro O. S. Alencar[3]

[1] Federal University of Pernambuco, Recife, Brazil
robertatdesa@gmail.com
[2] Military Fire Department of Pernambuco, Recife, Brazil
[3] Lajeiro Gesso Ltda, Recife, Brazil

Abstract. The paper deals with the fire resistance of non-structural walls made of solid plaster blocks. Experimental tests were performed according to the indications of Brazilian standard NBR 10636 [1], with a heat exposure time of 120 min, in three walls with thicknesses of 5, 7 and 10 cm. The results indicate that the plaster walls have great fire resistance: the maximum temperature measured on the unexposed face was 75 °C, occurred in the 5 cm thick sample. The 5 cm wall does not comply with the integrity and structural stability tests required by the above-mentioned standard during the heat exposure time, therefore, it is assumed that it can resist a fire exposure time of 90 min or less. The walls 7 and 10 cm thick comply with all criteria required by standard (thermal insulation, integrity and structural stability) during the heat exposure time, therefore, it can resist a fire exposure time of 120 min or higher.

Keywords: Fire safety · Heat resistance · Non-structural walls · Plaster walls

1 Introduction

Fires are tragic events, they can destroy patrimony and life. According to the International Association of Fire and Rescue Services (CTIF), about one-third of all recorded fire occurrences worldwide are in buildings [2], places of significant economic value and high concentration of people.

Fire safety is a science dedicated to study the most varied topics related to this problem. Among these studies, the resistance to fire of walls has been gaining prominence, since these elements compartmentalize environments and protects the structural elements, whether the wall is structural or non-structural. Walls are largely responsible for reducing the damage caused by fires occurred in buildings since they have appropriate resistance.

Due to the low cost, ease and agility of installation, the plaster walls are widely used in Brazil as compartmental systems for offices, shops, residences, among others, being present in much of current buildings.

P. A. G. Piloto et al. (Eds.): CILASCI 2019, LNCE 1, pp. 121–130, 2020.
https://doi.org/10.1007/978-3-030-36240-9_9

According to the US Geological Survey (USGS), Brazil is one of the world's largest producers of plaster [3], and only in the Araripe Region, located in the state of Pernambuco, accounts for 95% of national production [4], promoting local economy and development.

Given the importance of plaster wall resistance if fire situation, this paper aims to experimentally analyze their performance regarding thermal insulation, integrity and structural stability of non-structural walls made of solid plaster blocks, following the Brazilian standard NBR 10636 [1] requirements.

2 Experimental Program

2.1 Materials

Three walls with dimensions of 150 cm length and 150 cm height, and thicknesses of 5, 7 and 10 cm, respectively named W05, W07 and W10, were analyzed. These samples were made of solid plaster blocks with dimensions of 70 cm length and 50 cm height, with male-female fitting, connected with industrialized plaster mortar. None of the samples were coated and, after manufacture, all were stored in the laboratory free of wind and rain, remaining at room temperature (around 30 °C) for, at least, 30 days.

For the experiments, an electric furnace with a usable area of 1 m^2 was used (see Fig. 1a), programmed to reproduce the standard fire curve ISO 834-1 [5]. For temperature measurement, 10 K-type thermocouples were uniformly arranged inside the furnace and outside the furnace (unexposed face). The thermocouples inside the furnace remained at a distance of 10 cm from the exposed face of the walls, while the thermocouples on the unexposed face remained in direct contact with the plaster blocks, whose position did not coincide with the settlement mortar. The schematic disposition of the thermocouples inside and outside the furnace is shown on Fig. 1b.

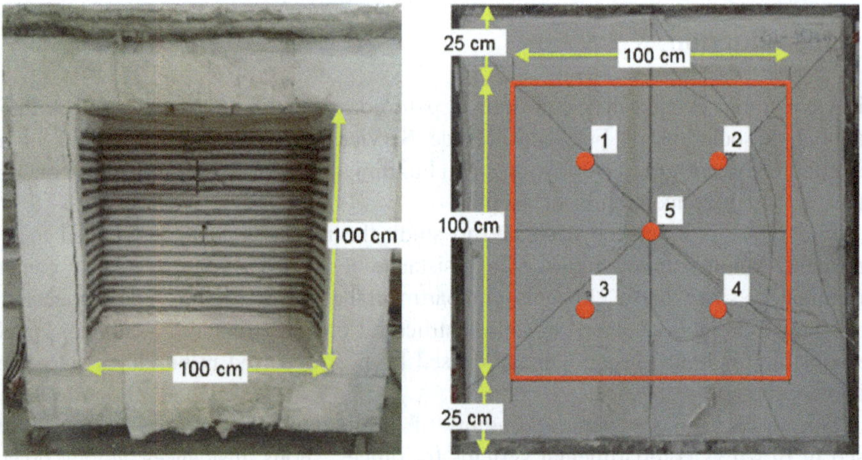

a. Electric furnace. b. Thermocouples inside the furnace and on the unexposed face to heat.

Fig. 1. Furnace and thermocouples location

In all tests, the temperature inside the furnace was defined by the arithmetic average of the temperatures measured by the thermocouples disposed at 10 cm from the exposed face to heat. While thermocouples outside the furnace provided the average temperature and maximum temperature on the unexposed face of the walls.

All tests were performed inside the Structure Laboratory of the Federal University of Pernambuco (Recife/Brazil), with pressure, humidity and air velocity under room conditions.

2.2 Furnace Heating

According to NBR 10636 [1], the temperature increase inside the furnace must be controlled to reproduce the standard fire curve, represented by the expression:

$$T = T_0 + 345log(8t + 1) \tag{1}$$

Where: T is the temperature at time t, given in °C; T_0 is the initial furnace temperature (room temperature), given in °C, where 10 °C $\leq T_0 \leq$ 40 °C; t is the test time, given in min. For this study, the room temperature was 30 °C and the test time was 120 min.

To ensure standardization of tests, NBR 10636 [1] states that the furnace temperature deviation shall be calculated as a percentage of the absolute error of the areas under the average furnace temperature curves and standard fire curve, establishing the following tolerances: (I) Before 10 min: ±15%; (II) Before 30 min: ±10%; (III) After 30 min: ±5%; (IV) From 10 min: the temperature measured by any thermocouple shall not differ by more than 100 °C from the corresponding standard temperature.

2.3 Test Methods

According to NBR 10636 [1], the following criteria should be evaluated during the test to determine the fire resistance of the walls:

Thermal Insulation. Characterized by the resistance of the wall to heat transmission, from the exposed to unexposed face to fire, preventing temperatures on the unexposed face from exceeding the following limits: (I) maximum average temperature at 140 °C; (II) an increase of 180 °C in any thermocouple with respect to room temperature, i.e. T_0 + 180 °C. The fail occurs when any of these criteria is met, indicating that the wall no longer resists to thermal insulation.

Integrity. Characterized by absence of sufficient openings (cracks) to allow the passage of hot gases or flames, from the exposed to unexposed face to fire. The presence of these openings is identified by the ignition test of a cotton wad.

Structural Stability. Characterized by the ability of the wall to remain intact, without collapsing, during all test period and when subjected to the mechanical shock test on the unexposed face. This test is performed with the impact of a 15 to 25 kg steel ball, in a pendulum motion, on the unexposed face to fire. On this study, the impacts were applied at three distinct points aligned horizontally at 75 cm height of the samples.

3 Results and Discussions

3.1 Furnace

For all tests, the tolerances required by NBR 10636 [1] were reached (see Table 1), showing that the furnace used is sufficiently reliable to guarantee the application of the thermal load required for such experiments. The Fig. 2 presents the temperature variation with time inside the furnace. Figure 2a shows the standard fire curve and the average temperature measured inside the furnace for sample test W05. In Fig. 2b it is also plotted the average furnace temperature curves for the three samples tested.

Table 1. Furnace temperature tolerances according to NBR 10636 [1].

	Error (%)			Maximum difference from 10 min (°C)	
	Before 10 min	Before 30 min	After 30 min		
Tolerance	±15	±10	±5	100	
W05	6.4	5.0	0.2	44	✓
W07	11.3	9.0	3.4	81	✓
W10	2.9	4.1	0.5	58	✓

✓ agree ✗ don't agree

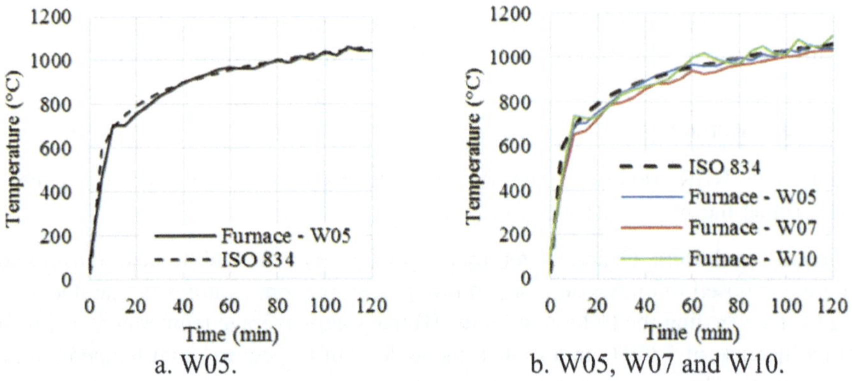

Fig. 2. Average temperatures inside the furnace.

3.2 Thermal Insulation

During the 120 min tests, all samples met the thermal insulation criteria required by NBR 10636 [1]. Figures 3a, 4a and 5a show the temperatures measured by thermocouples (T01, T02. T03, T04 and T05) on the unexposed face of samples W05, W07 and W10, respectively. Figures 3b, 4b and 5b show the average temperatures on the unexposed face of these same samples compared to their respective temperatures inside the furnace.

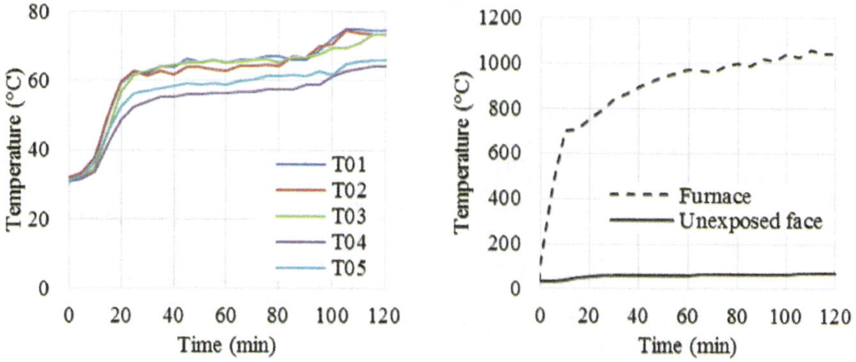

a. Temperature on the unexposed face

b. Average temperatures: inside the furnace and on unexposed face

Fig. 3. Temperatures measured on the unexposed face – sample W05.

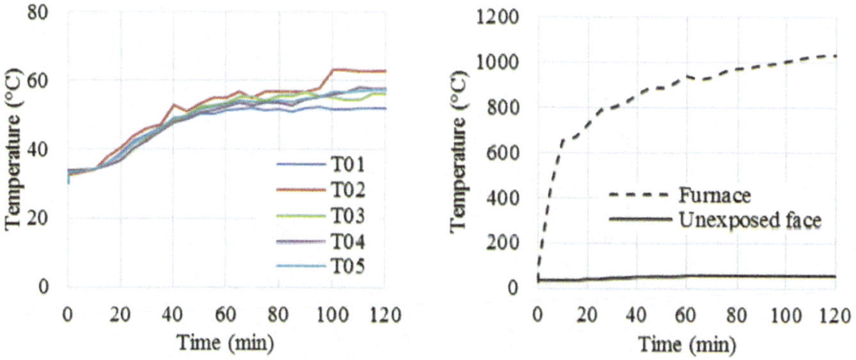

a. Temperature on the unexposed face

b. Average temperatures: inside the furnace and on unexposed face

Fig. 4. Temperatures measured on the unexposed face – sample W07.

As shown, none of the thermocouples attached to the unexposed face of the samples measured temperatures above 75 °C and therefore none of the average temperatures exceeded 70 °C. The maximum temperature measured in unexposed face was: W05: 75 °C; W07: 63 °C; W10: 55 °C.

Figure 6 shows the average temperatures on faces unexposed to fire in the tests. As expected, the sample W05 presented the highest average temperatures, presenting a difference of 26 °C from W10 at 25 min. W07 and W10 showed a maximum average temperature difference of 7 °C at 30 min. This shows that the 3 cm increment in W10 relative to W07 was not as effective as the 2 cm increment in W07 relative to W05.

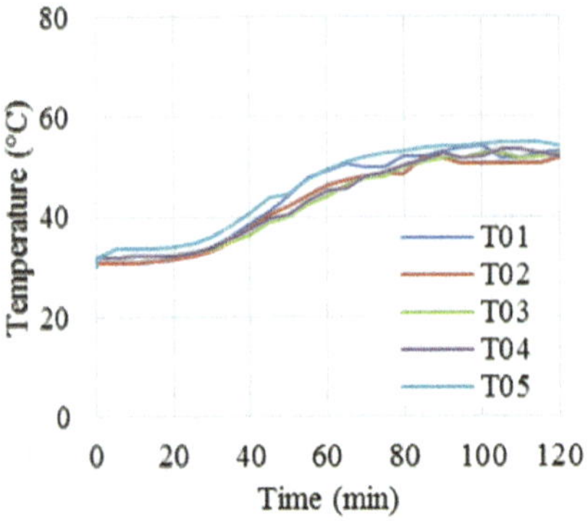

a. Temperature on the unexposed face

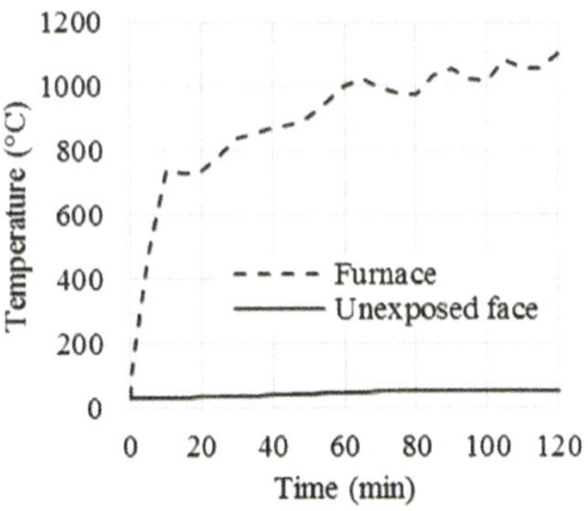

b. Average temperatures: inside the furnace and on unexposed face

Fig. 5. Temperatures measured on the unexposed face – sample W10.

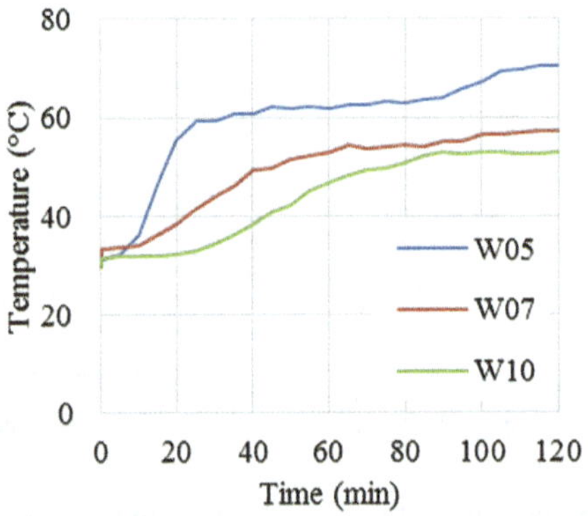

Fig. 6. Average temperatures on unexposed face of all samples.

3.3 Integrity

During all tests, no smoke or toxic gas emissions were identified; only the emission of a considerable volume of water vapors, due to the dehydration of the plaster, was observed. Moreover, it was observed that the appearance of cracks on the unexposed face of the walls occurred mainly in the connection of the plaster blocks, being precisely these places the most susceptible to the integrity loss in the samples.

On exposed face to fire, after the tests, it was observed that, besides the cracks at the plater blocks connection, a large part of the surface of the walls presented *hair cracks* (See Fig. 7), resulting from high dehydration and retraction of the plaster.

During the 120 min tests, the cracks at the unexposed face were monitored, and the ignition test of a cotton wad was performed periodically. Only W05 has lost its integrity ability and a cotton wad was ignited. This occurred at 115 min of test.

Fig. 7. Cracks on exposed face – sample W05.

No ignition of cotton wad occurred during the other samples' tests, preserving their integrity property up to 120 min tests.

3.4 Structural Stability

Throughout all tests, no sample showed any evidence of risk of collapsing without mechanical shock. The mechanical shock test was applied only at the end of the tests, i.e. at 120 min. Figure 8 shows the mechanical shock test of sample W05.

Clearly W05 collapsed due to the test. This occurred on the second impact. W07 and W10 did not have the same behavior, remaining stable even after the execution of the three impacts.

Another point observed was that none of the samples presented horizontal displacement that could be visually noticed, a fact clearly observed in fire resistance tests of ceramic and concrete block walls [6–10].

3.5 Classification

According to NBR 10636 [1], depending on how long the wall meets each criterion (thermal insulation, integrity and structural stability), they can fit in fire safety categories: Cut Fire Wall (Parede Corta-Fogo, CF) or Flame Arrester Wall (Parede Para-Chamas, PC). CF are those walls that attend all criteria during fire exposure and PC are those walls that attend only the requirements of structure stability and integrity.

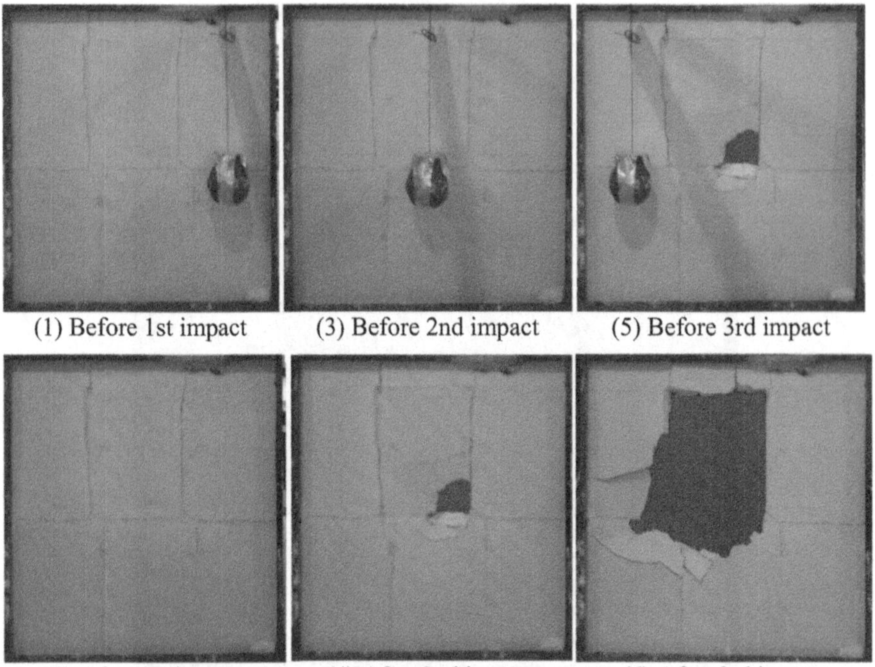

(1) Before 1st impact (3) Before 2nd impact (5) Before 3rd impact

(2) After 1st impact (4) After 2nd impact (6) After 3rd impact

a. 25cm from the right edge b. 75cm from the right edge (at the center) c. 125cm from the right edge

Fig. 8. Mechanical shock test – sample W05.

Thus, for correct classification of a wall, fire exposure times should be established, and the criteria required by NBR 10636 [1] must be met during the test. For example, for a wall to be classified as CF 120, it must meet the criteria of thermal insulation, integrity, and structural stability for 120 min of exposure to the standard fire curve. For a wall to be classified as PC 120, it must meet the criteria of integrity and structural stability for 120 min of exposure to the standard fire curve.

Table 2 presents a summary of the tests and the classification of the walls according to NBR 10636 [1].

Table 2. Summary of tests in plaster walls with heat exposure time of 120 min.

	W05	W07	W10
Thermal insulation	✓	✓	✓
Integrity	✗	✓	✓
Structural stability	✗	✓	✓
Classification	PC 90, CF 90 (*estimative*)	CF 120	CF 120

✓ requirement met ✗ requirement NOT met

For 120 min fire exposure, W05 wall met only the thermal insulation criterion, it did not resist the mechanical shock test and, at 115 min lost its the integrity property. Despite this, these results cannot be understood as conclusive, because the minimum requirement (integrity and structural stability) must be met throughout the fire exposure period predefined. For correct classification of this wall, fire exposure tests should be performed for 90 min (or another period). So, this classification PC 90, CF 90 for Wall 05 is only an *estimative*.

W07 and W10 met all the required criteria during the fire exposure time. These results can be understood conclusively for a CF 120 rating. However, with the results present here, it is not possible to affirm that walls W07 or W10 could not fit in a higher rating (PC 240, CF 240, PC 360 or CF 360, for example) because it would require more time of exposition to fire.

4 Conclusions

This paper presents the results of an experimental fire resistance study of nonstructural walls made of solid plaster blocks, following de requirements indicated by NBR 10636 [1]. For this, three samples with 5, 7 and 10 cm thickness were tested under the action of the standard fire curve for 120 min, allowing the following conclusions:

- The three walls tested were able to maintain the thermal insulation criterion during 120 min of fire exposure;
- The appearance of cracks on the unexposed face of the walls occurred mainly in the connections of solid plaster blocks;
- At 115 min, the 5 cm thick sample lost its integrity. This occurred in one of the connections of solid plaster blocks. 7 and 10 cm thick walls have not lost integrity throughout the fire exposure time;
- Only the 5 cm thick wall collapsed due to the mechanical shock test;
- The plaster walls presented good thermal performance, and the maximum temperature measured on the unexposed face was 75 °C, measured in the 5 cm thick sample;
- To the authors knowledge, it is not yet available, at the technical literature, similar results for wall of plaster blocks produced in Brazil; therefore more studies are still needed.

Acknowledgment. The authors are grateful for the research support granted by the Coordenação de Aperfeiçoamento de Pessoal de Nível Superior (CAPES), the Conselho Nacional de Pesquisa e Desenvolvimento Científico (CNPq), the Federal University of Pernambuco (Recife/Brazil), and are also grateful to the company Lajeiro Gesso Ltda that supplied the plaster blocks and the wall mounting.

References

1. Associação Brasileira de Normas Técnicas – ABNT: NBR 10636: Paredes divisórias sem função estrutural - Determinação da resistência ao fogo, Rio de Janeiro (1989)
2. International Association of Fire and Rescue Services – CTIF: Word Fire Statistics N°24. Center of Fire Statistics (CFS), National committees CTIF of Russia, Germany, USA (2019)
3. U.S. Geological Survey – USGS: Mineral commodity summaries 2019. Department of the Interior, U.S. Geological Survey (2019)
4. Federação das Indústrias do Estado de Pernambuco – FIEPE: Estudo Técnico - Polo Gesseiro do Araripe (2017)
5. International Standard – ISO 834-1: Fire-resistance tests—Elements of building construction —Part 1: General requirements, Switzerland (1999)
6. Rosemann, F.: Resistência ao fogo de paredes de alvenaria estrutural de blocos cerâmicos pelo critério de isolamento térmico. Master's dissertation in Civil Engineering, Technology Center, Federal University of Santa Catarina, Florianópolis, Brazil (2011)
7. Rigão, A.O.: Comportamento de pequenas paredes de alvenaria estrutural frente a altas temperaturas. Master's dissertation in Civil Engineering, Technology Center, Federal University of Santa Maria, Santa Maria, Brazil (2012)
8. Nguyen, Th.-D., Meftah, F.: Behavior of clay hollow-brick masonry walls during fire. Part 1: experimental analysis. Fire Saf. J. **52**, 55–64 (2012)
9. Borges, I.A., Rêgo Silva, J.J., Pires, T.A.C.: Alvenarias de vedação em blocos vazados de concreto simples submetidas a elevadas temperaturas – resultados preliminares. In: 4° Congresso Ibero Latino Americano sobre Segurança Contra Incêndio (4° CILASCI), pp. 33–42 (2017)
10. Borges, I.A.: Alvenarias de vedação em blocos de concreto simples submetidas a elevadas temperaturas características de processos de incêndio. Master's dissertation in Civil Engineering, Federal University of Pernambuco, Recife, Brazil (2018)

Author Index

P. A. G. Piloto et al. (Eds.): CILASCI 2019, LNCE 1, p. 131, 2020.
https://doi.org/10.1007/978-3-030-36240-9

The manufacturer's authorised representative in the EU is Springer Nature Customer Service Centre GmbH, Europaplatz 3, 69115 Heidelberg, Germany. If you have any concerns regarding our products, please contact ProductSafety@springernature.com

Printed and bound by CPI Group (UK) Ltd, Croydon, CR0 4YY
30/09/2026
02225259-0004